SpringerBriefs in Finance

More information about this series at http://www.springer.com/series/10282

Jorge M. Uribe · Montserrat Guillen

Quantile Regression for Cross-Sectional and Time Series Data

Applications in Energy Markets Using R

Springer

Jorge M. Uribe
Faculty of Economics and Business
Open University of Catalonia
Barcelona, Spain

Montserrat Guillen
Department of Econometrics
University of Barcelona
Barcelona, Spain

ISSN 2193-1720 ISSN 2193-1739 (electronic)
SpringerBriefs in Finance
ISBN 978-3-030-44503-4 ISBN 978-3-030-44504-1 (eBook)
https://doi.org/10.1007/978-3-030-44504-1

This Springer imprint is published by the registered company Springer Nature Switzerland AG
The registered company address is: Gewerbestrasse 11, 6330 Cham, Switzerland

Preface

This book project started during the writing of the Ph.D. thesis of Jorge, under the supervision of Montserrat, at the University of Barcelona in 2015. Quantile regression back then quickly became a useful tool to answer some intriguing research questions in economics and finance for several of the thesis chapters, and also as a preliminary data visualization device that served us to construct a complete panorama of the variables involved in the different economic models explored during the conducting of several research projects. We also noticed then that there was a lack of bibliographical resources approaching the topic from a rather eclectic perspective, which leaves out the technicalities that feature the methodology, and that focus on the more practical aspects of quantile regression implementation, and more importantly its interpretation. With this short-practical guide, we aim to fill at least partially this gap in the literature. This book is intended to a nonspecialized audience that might find useful the adoption of quantile regression as part of its empirical toolkit, useful to assess a large set of problems in many diverse settings within the disciplines of economics and financial economics. Indeed, we think that this book will certainly be useful for practitioners and academics with noneconometrics' previous exposure in different branches of the social sciences and engineering fields.

We wish to express our gratitude to Philipp Baun and Irene Barrios-Kezic, for their excellent editorial guidance at Springer. We also thank our colleagues at Riskcenter IREA, University of Barcelona, with a special mention to Albert Pitarque and Jessica Pesántez who helped us with code and text checking. BBVA Foundation grants on Big Data research and ICREA Academia are acknowledged.

Barcelona, Spain
February 2020

Jorge M. Uribe
Montserrat Guillen

About This Book

Quantile regression is an accurate and flexible econometric tool, useful to tackle some of the most common problems encountered in modern economics and finance, as for instance, measuring the effects of different environmental and market factors on production and consumption decisions made on a day-to-day basis by people and firms, or the modeling of series of prices in time, according to a market state or to different market seasonality. In what follows, we describe the use and interpretation of quantile regression, by modeling two different problems in energy markets: identification of the factors affecting households' energy consumption decisions and modeling the price in electricity markets. Energy markets will serve us a laboratory to carry out an empirical analysis using quantile regression as our workhorse for two different dimensions found in empirical applications: cross-sectional and time series data.

Contents

Chapter 1
Why and When Should Quantile Regression Be Used?

Abstract Quantile regression is a way to disclose predictive relationships between a response variable and some regressors or explanatory variables when the interest is to find a causal link beyond the mean-to-mean effects. Quantile regression is a procedure to model the cut points of the cumulated conditional probability distribution of a response variable as a function of some covariates.

Keywords Linear model · Regression · Quantile · Percentile · Predictive model · Risk analysis · Applied quantile regression

Since Koenker and Basset's (1978) seminal contribution, quantile regression models have gained significant importance for many fields of economics and beyond. Applications of quantile regression techniques in economics include recent studies on finance as in Fattouh et al. (2005) or Behr (2010); classical applications to modeling wages in labor economics (Chamberlain 1994; Buchinsky 1994); issues on the economics of education (Eide and Showalter 1998); inequality (Hao and Naiman 2007); modeling of prices and expenditure in different economic sectors such as traveling (Hung et al. 2010; Marrocu et al. 2015) or real estate (Zietz et al. 2008; Liao and Wang 2012); among other fields of economics—see also Koenker and Hallock (2001) and Fitzenberger et al. (2013). Outside economics, it has been used, for example, in genetics (Briollais and Durrieu 2014), medicine (Wei et al. 2006; Daniel-Spiegel et al. 2013), political science (Okada and Samreth 2012) or hydrology (Tareghian and Rasmussen 2013) to name some applications; see also Yu et al. (2003) for a recent summary of quantile regression research.

In sum, quantile regression enables the researcher to study the relationship between a set of variables not only at the center but also alongside the entire conditional distribution of the dependent variable.

In quantile regression models, the quantiles of a dependent variable are assumed to be linearly associated with a set of conditioning variables. In general, this translates into a nonlinear relationship between the dependent and the independent variables considering the whole distribution.

Quantile regression models constitute a promising tool for gaining a deeper understanding of financial markets and financial prices. In that context, the interest is not only on average prices, but also on volatility, i.e., on low or high extremes. Quantile

J. M. Uribe and M. Guillen, *Quantile Regression for Cross-Sectional and Time Series Data*, SpringerBriefs in Finance, https://doi.org/10.1007/978-3-030-44504-1_1

regression is a way to find what influences the magnitude of the response in those areas that are located far from the central value and that are not necessarily found in symmetric positions with respect to the mean.

Quantile regression models are known to be robust to outliers in the sample, which is particularly relevant when analyzing financial time series, in which crises and booms are generally accompanied by abnormally high or low observations. Quantile regression models are as well semi-parametric and, therefore, they require minimal distributional assumptions on the underlying data generating process, which allows a pragmatic approach when modeling complex dynamics as those recorded for financial prices and returns. Additionally, quantile regression models offer larger flexibility than linear regression models for analyzing different market scenarios. That is, lower quantiles of the return distributions in a given market can be naturally associated with downmarkets, while the higher quantiles are intuitively associated with up-trending markets. Therefore, very high or very low quantiles are likely related to the study of financial phenomena of great interest for the financial literature, such as bubbles, contagion or episodes of financial distress, which are known precisely for emerging under extreme market situations.

In the same way, quantile regression models are particularly convenient over competing linear regression alternatives, when the error structure is rather heterogeneous, also as is the case of many economic or financial time series, and when errors are not well described by a Gaussian distribution. As we will see in the following chapters, conditional quantile regressions also offer an intuitive and convenient way to model asymmetries and different regimes in the model for time series such as electricity prices.

In this book, we present quantile regression from a practical perspective. We show how it can be implemented for cross-sectional data and for time series data [see Alexander et al. (2011) for a perspective combining both times series and cross-sectional data]. We provide a comprehensive overview of the interpretability of the results. For this purpose, we have chosen the context of energy markets, because it is an area where quantile regression can be easily implemented and understood. The same methodology can immediately be implemented to other areas of economics and finance, and even for a broader audience of social scientists and engineers.

The next chapter provides a minimum introduction to energy markets as a case of study for quantile regression implementation throughout the rest of the chapters. This introduction to energy economics can be skipped for readers who are already familiar with this field. However, it gives an overview of the problems that are addressed in many other areas, both from the perspective of a cross-sectional analysis and of a time series analysis.

Chapter 3 is a general presentation to the main concept of quantile estimation, quantile regression and the difference between conditional and unconditional quantile regression. This is an essential part for understanding the modeling approach. Chapter 4 shows the implementation of quantile regression in cross-sectional data. Data from a survey, including sample weights, are used to illustrate the methodology. Programs in R and detailed outputs are presented so that the reader can follow the interpretation of the results. Chapter 5 is devoted to quantile regression for time series. Again the implementation is illustrated using data for electricity prices in the domain of energy markets, both for electricity and natural gas markets.

Chapters 6 and 7 are dedicated to the presentation of advanced topics and recent research contribution that are helpful to validate the models or to generalize quantile regression to a variety of contexts.

Chapter 8 concludes with a series of ideas for implementation and further research in this area. A short bibliographic list is provided at the end.

In addition to the examples presented here, we have worked on other areas where quantile regression has been an excellent toolkit. For example, in Bel et al. (2015) we were able to find that some policy interventions that reduce the pollution at the city of Barcelona were only effective at extreme episodes. On average, the effect was negligible, but when looking only at days with extremely dangerous environmental measurements, the action reduced pollutants significantly. In another totally different example (Pérez-Marín et al. 2019), risky drivers, defined as those who drive above posted speed limits, are easily identified with quantile regression techniques. It is an intuitive and very easy-to-understand example. Consider a driver who mainly drives in urban areas, then identifying that his risky behavior outstands that of similar drivers must be done by taking into account his driving zone, i.e., he is mainly driving in a metropolitan area. If this particular risky driver is compared to drivers in nonurban areas, then he might be considered prudent and safe.

From a more financial perspective in Chuliá et al. (2017), we employ multivariate quantile models to measure the responses of six main Latin American stock markets to a shock in the US index. We document asymmetrical responses to the shocks depending on the conditioning quantile at which they are calculated, which allow us to propose diversification strategies, which are only feasible after analyzing the quantile responses.

In demography, quantile estimations also provide a fresh perspective. For instance, in our work about quantile trends in the UK population (Uribe et al. 2018), we offer a new perspective for modeling past trends in the quantiles of the life function. Thanks to the quantile perspective we enrich the ongoing discussion about human longevity extension by calculating specific improvements in the distribution of the survivorship function, across its full range, and not only at the central-age ranges.

Finally, we should add that quantiles are just simple statistical measures that take into account relative frequencies. They differ from averages. However, even if quantiles can be called risk measures and a quantile is called value at risk in many areas, such as finance, there are now many more risk measures to be contemplated (Belles-Sampera et al. 2014, 2016). This means that quantile regression is only one of the many statistical measures of the response variable that a researcher may be interested in modeling.

All data and R codes that are used in this book are accessible via the https://github.com/montserrat-guillen/quantile_regression on the Web site.

References

Alexander, M., Harding, M., & Lamarche, C. (2011). Quantile regression for time-series-cross-section data. *International Journal of Statistics and Management System, 6*(1–2), 47–72.

Behr, A. (2010). Quantile regression for robust bank efficiency score estimation. *European Journal of Operational Research, 200*(2), 568–581.

Bel, G., Bolancé, C., Guillen, M., & Rosell, J. (2015). The environmental effects of changing speed limits: A quantile regression approach. *Transportation Research Part D: Transport and Environment, 36,* 76–85.

Belles-Sampera, J., Guillen, M., & Santolino, M. (2014). Beyond value-at-risk: GlueVaR distortion risk measures. *Risk Analysis, 34*(1), 121–134.

Belles-Sampera, J., Guillen, M., & Santolino, M. (2016). The use of flexible quantile-based measures in risk assessment. *Communications in Statistics—Theory and Methods, 45*(6), 1670–1681.

Briollais, L., & Durrieu, G. (2014). Application of quantile regression to recent genetic and -omic studies. *Human Genetics, 133*(8), 951–966.

Buchinsky, M. (1994). Changes in the US wage structure 1963–1987: Application of quantile regression. *Econometrica, 62,* 405–458.

Chamberlain, G. (1994). Quantile regression, censoring, and the structure of wages. In *Advances in econometrics: Sixth world congress* (Vol. 2, pp. 171–209).

Chuliá, H., Guillen, M., & Uribe, J. M. (2017). Spillovers from the United States to Latin American and G7 stock markets: A VAR quantile analysis. *Emerging Markets Review, 31,* 32–46.

Daniel-Spiegel, E., Weiner, E., Yarom, I., Doveh, E., Friedman, P., Cohen, A., et al. (2013). Establishment of fetal biometric charts using quantile regression analysis. *Journal of Ultrasound in Medicine, 32*(1), 23–33.

Eide, E., & Showalter, M. H. (1998). The effect of school quality on student performance: A quantile regression approach. *Economics Letters, 58*(3), 345–350.

Fattouh, B., Scaramozzino, P., & Harris, L. (2005). Capital structure in South Korea: A quantile regression approach. *Journal of Development Economics, 76*(1), 231–250.

Fitzenberger, B., Koenker, R., & Machado, J. A. (Eds.). (2013). *Economic applications of quantile regression*. Heidelberg: Springer.

Hao, L., & Naiman, D. Q. (2007). *Quantile regression* (Vol. 149). Thousand Oaks, CA: Sage.

Hung, W. T., Shang, J. K., & Wang, F. C. (2010). Pricing determinants in the hotel industry: Quantile regression analysis. *International Journal of Hospitality Management, 29*(3), 378–384.

Koenker, R. W., & Bassett, G. (1978). Regression quantiles. *Econometrica, 46,* 33–50.

Koenker, R., & Hallock, K. F. (2001). Quantile regression. *Journal of Economic Perspectives, 15*(4), 143–156.

Liao, W. C., & Wang, X. (2012). Hedonic house prices and spatial quantile regression. *Journal of Housing Economics, 21*(1), 16–27.

Marrocu, E., Paci, R., & Zara, A. (2015). Micro-economic determinants of tourist expenditure: A quantile regression approach. *Tourism Management, 50,* 13–30.

Okada, K., & Samreth, S. (2012). The effect of foreign aid on corruption: A quantile regression approach. *Economics Letters, 115*(2), 240–243.

Pérez-Marín, A. M., Guillen, M., Alcañiz, M., & Bermúdez, L. (2019). Quantile regression with telematics information to assess the risk of driving above the posted speed limit. *Risks, 7*(3), 80.

Tareghian, R., & Rasmussen, P. F. (2013). Statistical downscaling of precipitation using quantile regression. *Journal of Hydrology, 487,* 122–135.

Uribe, J. M., Chuliá, H., & Guillen, M. (2018). Trends in the quantiles of the life table survivorship function. *European Journal of Population, 34*(5), 793–817.

Wei, Y., Pere, A., Koenker, R., & He, X. (2006). Quantile regression methods for reference growth charts. *Statistics in Medicine, 25*(8), 1369–1382.

Yu, K., Lu, Z., & Stander, J. (2003). Quantile regression: Applications and current research areas. *Journal of the Royal Statistical Society: Series D (The Statistician), 52*(3), 331–350.
Zietz, J., Zietz, E. N., & Sirmans, G. S. (2008). Determinants of house prices: A quantile regression approach. *The Journal of Real Estate Finance and Economics, 37*(4), 317–333.

Chapter 2
A Case Study: Modeling Energy Markets by the Means of Quantile Regression

Abstract Quantile regression is a potent tool to analyze frequently found issues in economics and finance, such as the identification of consumption and production determinants and their potential impacts on demand and supply decisions, or the dynamics of prices that are featured by seasonality and other stylized facts that complicate traditional empirical modeling.

Keywords Energy markets · Energy prices · Extremes · Energy consumption · Energy policy · Covariates

Energy markets will serve us to illustrate our main points and to present the details of the implementation and interpretation of the modeling options that quantile regression has to offer us in these cases.

2.1 Energy Markets

Nowadays, electricity is considered a strategic asset that is determinant for virtually all sectors in modern production economies, in which it is a main input for most activities. Therefore, energy markets, in charge of the energy supply that guarantees the satisfaction of an ever-increasing world demand for consumption goods coming from a growing global population, have received considerable and increasing attention by the academic literature. In addition, from a purely financial perspective, investment in energy commodities such as oil has become a major mean to fostering financial diversification in traditional stock and bond portfolios, especially during episodes of market crash. The amount of assets allocated to commodity indices increased from \$13 billion in July 2003 to \$317 billion in July 2008, during the global financial crisis (Irwin and Sanders 2012). Thus, as emphasized by Mosquera et al. (2017), correct decisions about exposure and hedging against market risk depend on an accurate assessment of electricity price dynamics. Both exposure and hedging are fundamental for electricity producers and consumers.

Joint to variable renewable generation sources, i.e., wind and solar energy, which are progressively more and more present in the supply mix of electricity markets

J. M. Uribe and M. Guillen, *Quantile Regression for Cross-Sectional and Time Series Data*, SpringerBriefs in Finance, https://doi.org/10.1007/978-3-030-44504-1_2

across the world, the most important fuels used to generate power are oil, natural gas and coal. These three are generally competitors in the production of electricity, while all of them act as substitutes in consumption. This situation leads their prices to be closely interviewed and to respond to common fundamental factors and market shocks in rather nonlinear manners.

In terms of the stylized facts that characterize the series of prices in energy markets, the empirical literature has shown that electricity prices are featured by several facts, such as mean reversion, spikes, seasonal patterns, and long-range memory, which makes modeling and forecasting these prices a challenge for academics and practitioners in the electricity industry. The importance of the exercise of analyzing electricity prices is stressed out in both economic and financial terms, because the decisions on selling and buying are crucial for producers and consumers in energy markets and therefore for the rest of the economy.

In what follows we introduce quantile regression as a simple and intuitive tool that serves the purpose of modeling the consumption level and the price of electricity. When looking at time series of electricity price, quantile regression takes into account the characteristics of electricity prices, such as seasonality and heteroskedasticity in a very natural way, and as such it serves as a good starting point to understand the advantages that the methodology has to offer above more traditional linear regression alternatives. As we will see, using quantile regression to model energy prices also allows us to focus on specific fragments of the price distribution of electricity and to directly access the risk faced by either consumers or producers of electricity. Before actually seeing trends, we will examine consumption patterns by analyzing survey data in a cross-sectional analysis. The analysis reveals the heterogeneous nature of electricity consumption for a sample of households, showing that some factors affecting consumption have a substantially different impact depending on whether we are interested on mean (linear regression) or extreme (quantile regression) consumption, as a function of some covariates describing the household.

2.2 Energy and Quantile Regression: An Overview of Existing Analysis

Quantile regression provides a natural tool to tackle some of the most relevant research questions in the field of energy markets, and as such it has been employed by previous studies in the literature to analyze a variety of issues, some of which we summarize in what follows. This set of studies constitutes a starting point for the analysis that we conduct, and they will hopefully provide the readers with the motivation and perspective required to fully take advantage of the tools presented herein.

One of the most natural applications of quantile regression emerges in the field of energy policy. Particularly, quantile regression has proven to be a potent tool when it comes to analyze the spectrum of households' energy consumption. Kaza (2010) states that the residential sector in the USA accounts for the 22% of the annual energy consumption, following closely the share of the transportation sector with a 28%. Hence, the study of energy demand on the side of households becomes of first-order importance for regulators and planners engaged in the market of energy, especially when such regulators attempt to push forward in the political agenda energy conservation polices. Indeed, understanding the effects of socioeconomic and housing characteristics on the consumption of energy is relevant for groups interested in influencing pro-environmental and conservation policy.

With this in mind, Kaza (2010) explores the effects of housing characteristics such as density, type and location on the patterns of energy consumption. These same variables have been consistently used to explore the average energy demand of households, but as the author of the study emphasizes, quantile regression offers a unique opportunity to analyze the whole conditional distribution of the residential energy consumption, and by so doing, to determine whether the tail effects and the effects on other parts of the consumption distribution differ from the average effect. If the measured effects of the explanatory variables turn out to be different at different parts of the energy consumption distribution, then design of public polices seeking to reduce energy consumption might be targeted directly to reduce it at the higher end of the energy consumption spectrum. In sum, the study finds that household size matters for cooling energy uses, housing size matters for all energy uses and neighborhood density does not affect the energy use directly. The author also documents an important role of climate variables at explaining energy consumption. Owners compare to renters, use less cooling energy but use more other energy. Furthermore, the marginal impact of multifamily units on large apartment buildings compared to single-family detached houses is significantly different across the spectrum of energy consumption and, therefore, relying on average effects for the designing of public policies, undercuts the savings that can be gained from tailoring policies to suit large energy consumption households.

Valenzuela et al. (2014) also study households' energy consumption patterns by the means of conditional quantile regressions. The authors examine energy consumption from single-family households as a function of household demographics and housing unit characteristics. The authors emphasize that modeling household energy consumption is a difficult task due to the large variability of behaviors, which in turn contributes to observed differences in consumption patterns, even among households inhabiting similar housing units and which share demographic characteristics. Quantile regression models allow the authors to compare how various quantiles of energy consumption are affected by changes in the predictor characteristics in relation to other quantiles. The changes in the magnitude and direction of the regression coefficients for a given explanatory variable from one quantile to another suggest differences in energy consumption that might be more affected by the predictor variable at one part of the energy consumption distribution in comparison to another part of the distribution.

The results of this study suggest that targeting households that have pools, non-central cooling, pier/post foundations and/or are renter-occupied may result in advancing energy conservation efforts. Policy implications from this study also suggest that demographic characteristics of household composition and daily patterns of unit occupation are important aspects of consumer demand for energy. The authors provide supportive evidence on different behavioral patterns due to life-cycle stages and transitions or generational differences. Overall, the results presented in the study show that there is a need to design strategies based on the location of households within the distribution of energy consumption. For instance, some of the main findings regard the bottom 10th quantile of energy consumers. Although such households are naturally the lowest consumers, it would be convenient for the utility firm to considering offering gas services to households without gas connections. On their side, regression results at the 90th quantile show that there is a need to reduce energy consumption among household without central AC units.

More recently, Niemierko et al. (2019) analyze residential heating energy consumption employing quantile regression and D-Vine copulas. The authors highlight that the "residential building sector is one of the biggest energy consumers globally and yet, while policy-makers have focused on the energy efficiency potential of residential buildings, the actual resulting investment volume in retrofit artifacts is disappointing." They claim that the reason for this is the so-called performance gap between predicted and realized energy savings. The gap appears because data-driven building energy models are mostly fitted on synthetic data generated by simulation software. Thus, over- and underestimation of energy consumption is a frequently observed phenomenon. That is, the simulation software neglects the variance in actual energy consumption, as, for example, that triggered by different behaviors of the building tenants. Data-driven approximations to the analysis of energy consumption, which are based on historical energy consumption and detailed information about the building, may well overcome these issues. However, frequently used point estimation methods, such as artificial neural networks, exhibit major drawbacks, whenever the variance in the data is high. On its side, quantile regression designs potentially overcome such issues by estimating different intervals instead of point (single) values.

The quantile regression design followed by the authors provides complete information on the distribution of heating energy consumption, which is conditional on a set of building parameters (e.g., building envelope situations, building age, heating systems, number of windows). They find that consumption sensitivities vary across quantiles, and therefore, the rebound effect can be analyzed by using their approach. Such rebound effect occurs whenever an increase in the demand of energy is observed after implementation of an energy efficiency measure, and the performance gap at different parts of the distribution of energy consumption. Their contribution to the literature is evident, because it allows decision-makers to better comprehend the efficacy of different energy retrofit measures depending on household and building properties. For instance, the quantile regression approach reveals that very energy-conscious and energy-wasteful households are prone to more extreme rebound effects, and regarding to the performance gap, it is found that heating energy demand of inefficient buildings is systemically underestimated, whereas efficient buildings are overestimated.

References

Irwin, S. H., & Sanders, D. R. (2012). Financialization and structural change in commodity futures markets. *Journal of Agricultural and Applied Economics, 44*(3), 371–396.

Kaza, N. (2010). Understanding the spectrum of residential energy consumption: A quantile regression approach. *Energy Policy, 38,* 6574–6585.

Mosquera, S., Manotas, D. F., & Uribe, J. M. (2017). Risk asymmetries in hydrothermal power generation markets. *Electric Power Systems Research, 147*, 154–164.

Niemierko, R., Toppel, J., & Trankler, T. (2019). A D-vine copula quantile regression approach for the prediction of residential heating energy consumption based on historical data. *Applied Energy, 233–234,* 691–708.

Valenzuela, C., Valencia, A., White, S., Jordan, J. A., Cano, S., Keating, J., et al. (2014). An analysis of monthly household energy consumption among single-family residences in Texas, 2010. *Energy Policy, 69,* 263–272.

References

Chapter 3
Quantile Regression: A Methodological Overview

Abstract In this chapter, the main methodological concepts related to quantile regression are described. We provide the definition of conditional and unconditional quantiles and present the minimization problem with asymmetric loss that underlies the quantile estimation via quantile regressions. Additionally, weighted quantile regression tools that will be used in the following sections of the book are presented at the end of this chapter.

Keywords Least squares · Maximum likelihood · Loss minimization · Median regression · Parameter estimation · Conditional quantile · Unconditional quantile

Quantile regression is a method that aims at fitting the quantile of the cumulative distribution of the response variable, for a fixed confidence level, given some covariates. So, a quantile of the response variable is expressed as a linear combination of the covariates, and estimating the model implies finding the coefficients for that linear combination. Results vary from one confidence level to another. For instance, for a confidence level of 50%, the 50th quantile regression corresponds to a model for the median of the conditional distribution of the response. At level 95%, a quantile regression is a model for the 95th quantile, i.e., the value of the conditional distribution of the response that only 5% cases should exceed. We need to say "conditional distribution," because the quantile is fitted for some given fixed values of the covariates.

The objective of quantile regression differs from the aims of classical linear regression. A linear regression model expresses the dependent variable as a linear combination of some fixed covariates plus a random error term. So, linear regression fits the mean of the response, because it is assumed that the error term has expectation equal to zero. Quantile regression studies the quantiles rather than the expectation of the response variable.

Quantile regression is a wonderful method to understand what influences the possibility of having extreme responses, because it is based on the notion of the cumulative distribution function. In this chapter, we define quantiles and we also show how to estimate a quantile in a univariate framework, as the minimum of an objective function that can be expressed as an expected loss. This minimization procedure is crucial to understand the estimation principle underlying quantile regression. There

J. M. Uribe and M. Guillen, *Quantile Regression for Cross-Sectional and Time Series Data*, SpringerBriefs in Finance, https://doi.org/10.1007/978-3-030-44504-1_3

are excellent presentations that introduce practitioners to quantile regression (see, e.g., Conde Amboage et al. 2018; Petscher et al. 2013).

Let X_i be the vector of covariates for individual *i*, and let Y_i be the dependent variable for individual *i*. Let us consider a parameter vector β with the same dimension as the vector of covariates, we express the linear combination as $X_i'\beta$, where $'$ denotes the transposed vector. Let ε_i be a random variable that corresponds to the *i*-th error term. In a linear regression setting:

$$Y_i = X_i'\beta + \varepsilon_i, \tag{3.1}$$

and assuming that $E(\varepsilon_i) = 0$, then $E(Y_i|X_i) = X_i'\beta$.

Parameter estimates can be obtained by the method of ordinary least squares (OLS). This means that parameter estimates are obtained by minimizing the sum of squared residuals:

$$\text{argmin}_\beta \sum\nolimits_{i=1}^{n} [y_i - E(Y_i|X_i)]^2 = \text{argmin}_\beta \sum\nolimits_{i=1}^{n} \left[y_i - X_i'\beta\right]^2, \tag{3.2}$$

where n is the total number of observations and the residuals are defined as the difference between the observed values of the response, y_i, which we denote with lowercase letters, and their corresponding conditional expected value, $E(Y_i|X_i)$.

If we assume that the response variable follows a normal distribution, then OLS estimates are equivalent to maximum likelihood (ML) estimates and the asymptotic properties of the ML method apply. For example, the estimators of the parameters are normally distributed asymptotically, and their standard errors can also be approximated. Confidence intervals and tests of hypothesis are easily derived.

We are interested in seeing what needs to be changed if we aim at modeling the quantile of the dependent variable and not its expectation.

3.1 Definition of Quantile and Conditional Quantile

Let us first define a *quantile in a univariate framework*. Let us assume that Y is a random variable with probability distribution function equal to $F(y)$, so $0 \leq F(y) \leq 1$ is by construction a nonnegative function. At level θ, $0 < \theta < 1$, the θ-th quantile of Y_i is defined as:

$$Q_Y(\theta) = \inf\{r|F(r) \geq \theta\}. \tag{3.3}$$

If we consider that the cumulative distribution function of individual *i*, $F_i(y)$, depends on the covariates, then the θ-th *conditional quantile* of Y_i given X_i is defined as:

$$Q_{Y_i|X_i}(\theta) = \inf\{r\text{: } F_i(r|X_i) \geq \theta\}. \tag{3.4}$$

3.2 Estimating the Quantile in the Univariate Case

Let us start from (3.3) in the univariate framework. We need to find a way to estimate $Q_Y(\theta)$, for a fixed level θ, now without covariates as we stay in the univariate case, and using only information on the dependent variable Y. Let us assume that a sample of n data points is observed: $y_1, y_2, \ldots, y_n$. There are several routes to estimate $Q_Y(\theta)$. First, we can assume that $F(y)$ is known, and it is strictly increasing and continuous. Then, we can find its inverse, either numerically or analytically, and so $Q_Y(\theta) = F^{-1}(\theta)$ is uniquely defined. This approach is convenient for well-known statistical distributions such as the normal, lognormal, Pareto and exponential. Second, if no assumption on $F(y)$ is possible, then we can estimate $F(y)$ using a nonparametric technique and then we can find the solution of the following equation, for instance, using a numerical method:

$$Q_Y(\theta) = \text{argmin}_r \left[\hat{F}(\text{r}) - \theta\right]^2,$$

where $\hat{F}(\text{r})$ denotes the estimated cumulative distribution function. This method is indirect, because it is based on an estimate of the cumulative distribution function $F(y)$.

The most celebrated method to estimate a quantile $Q_Y(\theta)$ is based on the following objective function:

$$Q_Y(\theta) = argmin_c E[\rho_\theta(Y - c)], \tag{3.5}$$

where $\rho_\theta(\cdot)$ is a loss function. The advantage of this approach is that the quantile is defined as a minimum expected loss. The loss function is as follows:

$$\rho_\theta(u) = \left[\theta - I_{\{u<0\}}\right]\text{u},$$

where $I_{\{u<0\}}$ is an indicator function that is equal to 1 when the subscript is true and 0 otherwise. The loss function $\rho_\theta(\cdot)$ is nondifferentiable. When $u < 0$, then it is a straight line with slope equal to $\theta - 1$, and when $u > 0$, then it equals a straight line with slope equal to θ. So it can be expressed as: $\rho_\theta(u) = (1-\theta)I_{\{u<0\}}|\text{u}| + \theta I_{\{u>0\}}|\text{u}|$. See Fig. 3.1.

Statistical measures that can be expressed as a minimum of a loss function are called elicitable (see Fissler and Ziegel 2016). Quantiles are, therefore, elicitable statistical tools.

When finding the quantile for a sample of data, the loss function is evaluated in (Y − c) for each possible c value. Our aim is to find c that minimizes the expected loss. So, the loss function weights positive deviations from c with the load equal to θ and negative deviations, i.e., $(Y - c) < 0$, with a loading equal to $\theta - 1$.

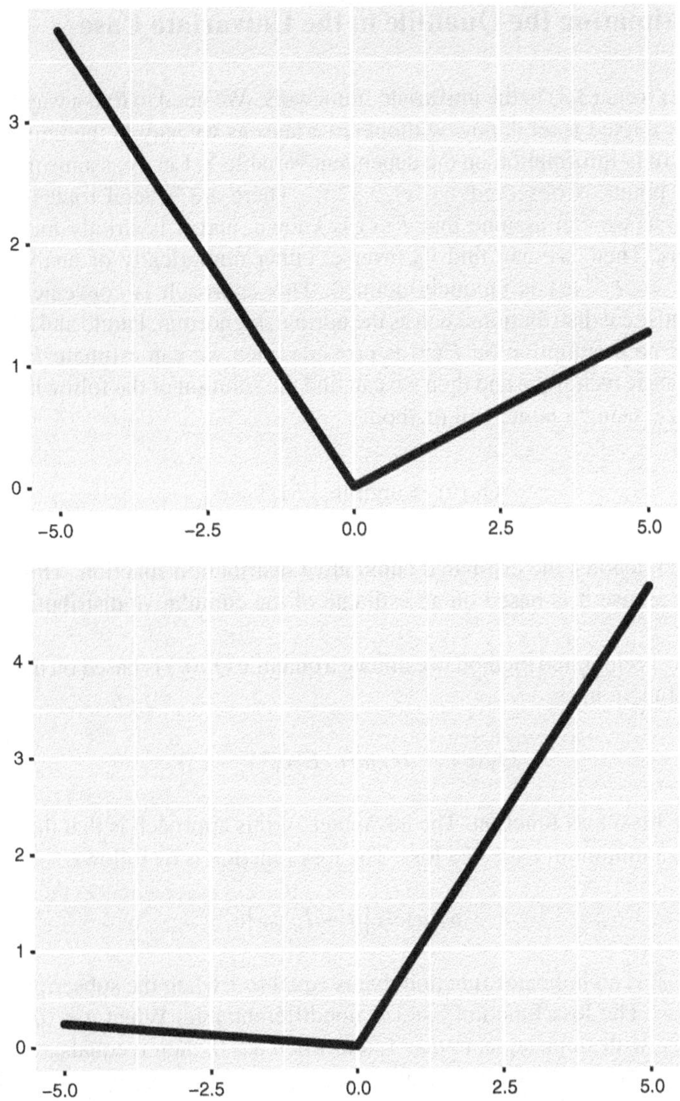

Fig. 3.1 $\rho_\theta(\cdot)$ is found here for $\theta = 0.25$ (top) and $\theta = 0.95$ (bottom). *Source* Illustration by the author

The proof of (3.5) is not so straightforward. If we assume that Y is a continuous random variable, with probability density function f(y), then the expected value of the absolute deviations from a given center c can be split into the following two terms:

$$E(|Y - c|) = \int |y - c| f(y)dy = \int_{-\infty}^{c} -(y - c)f(y)dy + \int_{c}^{+\infty} (y - c)f(y)dy.$$

To find the minimum expected loss, let us calculate the derivative with respect to c:

$$\frac{\partial}{\partial c} E[\rho_\theta(Y - c)] = \frac{\partial}{\partial c} \int_{-\infty}^{c} (1 - \theta)(c - y)f(y)dy + \frac{\partial}{\partial c} \int_{c}^{+\infty} \theta(y - c)f(y)dy.$$

Taking the derivative and integrating by parts leads to:

$$(1 - \theta)(c - y)f(y)|_{-\infty}^{c} +$$

$$\int_{-\infty}^{c} \frac{\partial}{\partial c}(1 - \theta)(c - y)f(y)dy + \theta(y - c)f(y)|_{-c}^{+\infty} + \int_{c}^{+\infty} \frac{\partial}{\partial c}\theta(y - c)f(y)dy$$

$$= 0 + (1 - \theta) \int_{-\infty}^{c} f(y)dy + 0 - \theta \int_{c}^{+\infty} f(y)dy$$

$$= \int_{-\infty}^{c} f(y)dy - \theta = F(c) - \theta.$$

So, when $\frac{\partial}{\partial c} E[\rho_\theta(Y - c)] = 0$, then $F(c) = \theta$, which is exactly the condition to be the θ-th quantile.

In the proof, we have used the standard properties of the probability density function, i.e., $\int_{-\infty}^{+\infty} f(y)dy = 1, \lim_{y \to \pm\infty} f(y) = 0$.

At this point, the fact that a quantile can be found as the minimum of the expectation of a loss function may seem a bit artificial. This is not the case. This property is known as being *elicitable*. If we go back to the point where we assumed that a sample of n data points is observed $y_1, y_2, \ldots, y_n$, then we can see the interest of this approach. If instead of the expectation, we consider the mean so that the expected loss $E[\rho_\theta(Y - c)]$ can be estimate with the sample average as:

$$\frac{1}{n} \sum_{i=1}^{n} \rho_\theta(y_i - c).$$

The equivalent expression using the definition of the loss function, $\rho_\theta(u) = (1 - \theta)I_{\{u<0\}}|u| + \theta I_{\{u<0\}}|u|$, is:

$$\frac{1}{n} \sum_{i=1}^{n} (v)I_{\{y_i<c\}}|y_i - c| + \theta I_{\{y_i>c\}}|y_i - c|. \tag{3.6}$$

So, when a sample is available, a numerical search suffices to find $Q_Y(\theta) = c$, so that (3.6) is minimum. If $\frac{1}{n} < \theta < \frac{n-1}{n}$, minimizing (3.6) with respect to c is always possible and there are fast algorithms to find the quantile. However, if there are no enough data, i.e., n is small, then for too high or too low quantiles, the estimate would be equal to the maximum or minimum observation.

3.3 Quantile Regression Estimation

When covariates are present, then estimating the conditional quantile $Q_{Y_i|X_i}(\theta)$ for the i-th observation is a question of fitting a parameter vector, such that the expected loss function is minimum. The procedure mimics the approach of classical linear regression where the vector of parameters is fitted by minimizing the sum of squared residuals, which in that case is a quadratic loss function.

So, in quantile regression, the aim is to find the vector of parameters, $\beta_{(\theta)}$, in a quantiles regression model, where

$$Q_{Y_i|X_i}(\theta) = X_i'\beta_{(\theta)}. \tag{3.7}$$

Note that the parameter vector has a superscript, because the vector changes for each value of the θ quantile level. So, going back to Expression (3.5), we have that:

$$\beta_{(\theta)} = argmin_{\beta} E\left[\rho_{\theta}\left(Y - X_i'\beta\right)\right], \tag{3.8}$$

with, as before,

$$\rho_{\theta}(u) = (1 - \theta) I_{\{u<0\}} |u| + \theta I_{\{u>0\}} |u|. \tag{3.9}$$

The solution to the median case ($\theta = 0.5$) is the line that minimizes the sum of absolute deviations, which is the sum of the length of the residuals.

The mathematical formulation of the problem leads to the solution of a linear programming problem. Its basic structure and the counterpart algorithm solution are omitted here. More information can be found in Davino et al. (2013) and Koenker (2005).

3.4 Quantile Regression Estimation Versus Weighted Quantile Regression Estimation

Let Y_i be the vector of the response variable that corresponds to the dependent variable for each observation *i*, and X_i is the vector for observation *i* of j covariates, i = 1, …, n and j = 1, …, K, where n equals the sample size and K denotes the number of covariates.

The conditional distribution function of Y_i given X_{ij} is $F(y|X_{ij}) = P(Y_i \leq y|X_{ij})$. For $0 < \theta < 1$, the θth conditional quantile of Y_i given X_i is defined as

$Q_{Y_i|X_i}(\theta)\inf\{r:\ F(r|X_{ij}) \geq \theta\}$. Then, the linear quantile regression assumes that $Q_{Y_i|X_i}(\theta) = X'_{ij}\beta_{(\theta)}$, where $\beta_{(\theta)}$ represents the vector of unknown parameters, and the model specification may be written as:

$$Y_i = X'_i\beta_{(\theta)} + u_i, \tag{3.10}$$

where u_i is the error term that corresponds to each observation and satisfies theoretically that $P(u_i < 0|X_{ij}) = 0$.

The θth quantile regression estimator $\hat{\beta}_{(\theta)}$ minimizes over $\beta_{(\theta)}$ the following objective function:

$$Q_{\beta_{(\theta)}} = \sum\nolimits_{i:Y_i \geq X'_i\beta_{(\theta)}}^{n} \theta\left|X'_i\beta_{(\theta)}\right| + \sum\nolimits_{i:Y_i < X'_i\beta_{(\theta)}}^{n} (1-\theta)\left|X'_i\beta_{(\theta)}\right|, \tag{3.11}$$

with $0 < \theta < 1$. In other words,

$$\hat{\beta}_{(\theta)} = \underset{\beta}{\operatorname{argmin}} \sum\nolimits_{i=1}^{n} \rho_\theta\left(Y_i - X'_i\beta_{(\theta)}\right), \tag{3.12}$$

And the weighted θth quantile regression is specified as:

$$\hat{\beta}_{(\theta)} = \underset{\beta}{\operatorname{argmin}} \sum\nolimits_{i=1}^{n} W_i\,\rho_\theta\left(Y_i - X'_i\beta_{(\theta)}\right), \tag{3.13}$$

where W_i represents the vector of weights, in this case corresponds to the variable NWEIGHT, and $\rho_\theta(u) = u\{\theta - I(u < 0)\}$ is the quantile loss function. Let us note that the previous two equations differ because of the vector of weights.

References

Conde Amboage, M., González Manteiga, W., & Sánchez Sellero, C. A. (2018). Quantile regression: Estimation and lack-of-fit tests. *BEIO, Boletín de Estadística e Investigación Operativa, 34*(2), 97–116.

Davino, C., Furno, M., & Vistocco, D. (2013). *Quantile regression: Theory and applications* (Vol. 988). New York: Wiley.

Fissler, T., & Ziegel, J. F. (2016). Higher order elicitability and Osband's principle. *The Annals of Statistics, 44*(4), 1680–1707.

Koenker, R. W. (2005). *Quantile regression*. Cambridge: Cambridge University Press.

Petscher, Y., Logan, J. A., & Zhou, C. (2013). Extending conditional means modeling: An introduction to quantile regression. In *Applied quantitative analysis in education and the social sciences* (pp. 15–45). London: Routledge.

Chapter 4
Cross-sectional Quantile Regression

Abstract In this chapter, we illustrate the classical problem of quantile regression for cross-sectional data. We develop one practical exercise using R. We visualize and interpret our results, where quantile regression is used to explain which factors affect excess consumption of electricity by a sample of US households with different characteristics. These results are helpful to evaluate the risk of disproportionate consumption. Our innovation here, rather than the method, is our practical approach. We fully develop our example, guiding the reader step by step from the previsualization of the data to the interpretation of the coefficients and the diagnostics tests to be performed before and after the estimation of the parameters.

Keywords Sample data · Sampling weights · Residential energy consumption survey · Parameter estimation · Inference · Interpretability

In what follows we illustrate quantile regression for cross-sectional data that were collected in a survey with nonproportional representation of the units. This means that not all households in our sample represent the same number of households in the population. This is a common problem of economic surveys, where sampling weights are necessary to extrapolate conclusions beyond to the surveyed population. In order to do a step-by-step presentation, we start by introducing our real dataset case on energy consumption that collects information on a sample of US households.

There are three challenges that we want to address in this chapter besides presenting how quantile regression is implemented in practice using R. First, we are going to work with survey data where each observation does not have the same importance in the sample, and in other words, each observation is associated with a sample weight which measures the number of actual households that are represented by each particular observed unit. All analyses that aim to create models for the population need to be using the weighting variable. Otherwise, the conclusions would only apply to the particular sample data and not to all the households of the US population of households. In other words, unweighted analysis means that data are a biased representation of the population of interest, so conclusions might also be biased if models do not control for the variables that are used in the sample design. For instance, if our models do not contain information of the geographical zone or demographics of the observed units, then we may be subject to substantial sampling bias; see Solon

J. M. Uribe and M. Guillen, *Quantile Regression for Cross-Sectional and Time Series Data*, SpringerBriefs in Finance, https://doi.org/10.1007/978-3-030-44504-1_4

et al. (2015). We encourage the reader to follow the statistical debate in Pfeffermann et al. (1998) for an informed decision on using sampling weights in correct statistical modeling. Second, we need to understand the difference in the interpretation of the coefficients of an OLS regression versus a quantile regression, and for this purpose, we start with a simple model with only one explanatory variable. Third, an additional section is devoted to implementing a more general cross-sectional model where quantile energy consumption is explained by a set of explanatory variables. We will then use visualization tools to interpret the results in this multivariate setting.

The case study of energy consumption provides a simple setting to present the main ideas for a straightforward implementation. Electricity is one of the best inventions in history worldwide; thus, human life without it will be almost impossible nowadays. The importance of electricity is found in several fields of economic activity such as engineering, communications, transport, medical care, among others. As a result, electricity consumption is increasing, and this fact has several economic and environmental effects such as an increase in utility bills that affects household savings, but also a strain of an already overtaxed power grid which can contribute to climate change. Then, energy-pricing policies are demanded to control negative effects of overconsumption. Studying the factors that influence the total residential electricity consumption may provide guidelines to adjust the demand of energy.

Apart from this general motivation, our analysis should consider that electricity pricing is generally determined by supply and demand, so it is necessary to analyze the impact of the factors on the distribution of energy consumption, not only on its mean because pricing thresholds could be established according to the quantiles of consumption. Quantile regression offers a convenient tool to established quantiles of consumption that vary depending on the household's characteristics. This allows establishing levels of consumption determined by quantiles that actually depend on facts like the number of family members, which may determine a fair price to be paid for overconsumption.

4.1 Data Source

The Residential Energy Consumption Survey (RECS) is an American sample poll that collects information related to energy consumption, expenditure and characteristics of housing units occupied as a primary residence and the households that live in them. The RECS is conducted every 6 years since 1978.

RECS 2015 which is used for this analysis has a sample of 5686 US households, selected randomly based on a complex multistage, area-probability sample design. RECS 2015 added sample cases that were weighted to represent the population so as to extend the analyses for the total 118.2 million US households. Therefore, a case's weight indicates the number of households that the particular case represents. The variable NWEIGHT represents the final sampling weight, which is the number of households in the whole population that the observation represents (see further details in RECS 2015 Technical Documentation—Summary). Consequently, our models aimed at finding results that are inferred for the whole US population of households incorporate the final sampling weights.

Table 4.1 describes the variables that will be used in our examples. We will use the variable *Energy Consumption* as our dependent variable, which measures the cost of electricity paid by each observed household in 2015 for the entire year. In the first model attempt, we just explain the total cost in year 2015 by the surface of the household. For illustrative purposes, we have also selected a small set of additional covariates in order to formulate a more general model in a later section.

First of all, it is mandatory to explore the distribution of the dependent variable which is the cost of electricity and represents a proxy variable for energy consumption of the households. Figure 4.1 shows this variable's quantiles plotted against the fraction of the data it corresponds to, sorted from the minimum to the maximum. The idea is to detect how the variable's quantiles behave. Therefore, there is a low cost of electricity for the lower quantiles and around 2000 US$ for the median cost with a smooth increase until the 90% quantile approximately. But after that, its distribution has a sharp peak.

The behavior of the cost of electricity seems to significantly differ for householders with lower quantiles of energy consumption from the higher ones. The first

Table 4.1 Variables of the RECS 2015 used in this chapter

Type	Variable name	Description
Dependent	Energy consumption	Total electricity cost in dollars, 2015
Independent	Swimming pool	Dummy variable equal to one if the household has swimming pool, and zero otherwise
	Square footage	Total square footage
	Cooled square footage	Total cooled square footage
	Heated square footage	Total heated square footage
	Bedrooms	Number of bedrooms
	Electricity generation from solar	Dummy variable equal to one if the household has on-site electricity generation from solar, and zero otherwise
	Num. TVs	Number of televisions used
	Smart phones	Number of smart phones
	Floor, window or table fans used	Number of floor, window or table fans used
	Household members	Number of household members
Weighting	NWEIGHT	Number of households that are represented by observation. It must be used to obtained unbiased estimates

Source Constructed by the author with information from RECS 2015

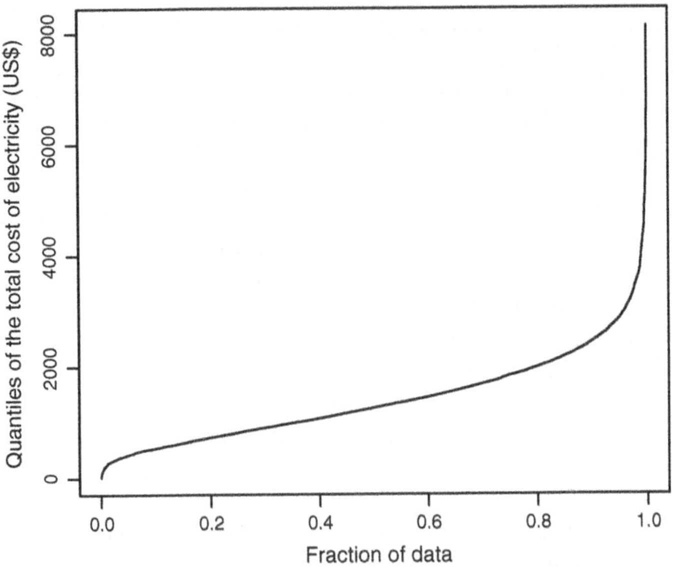

Fig. 4.1 Quantiles for cost of electricity (US$) from RECS in 2015. *Source* Illustration by the author

statistical motivation to apply quantile regression to fit these data is that the situation presented by RECS 2015 energy consumption suggests a different relationship between the exogenous variables and each conditional quantile of the endogenous variable rather than a link between the expected consumption and its mean as ordinary linear regression does. However, this decision will be validated later.

4.2 Weighted Versus Unweighted Linear Regression: A Simple Example

Before estimating any complicated model, we will describe the role of weights in a simple setting and then we will formally define linear regression for unweighted and for weighted data, before showing the differences between quantile regression and weighted quantile regression model.

Let us start with a classical regression. We will first explain energy consumption of one US household by the surface of the household's residence, but we will start by seeing the difference between using the weights and omitting them. Figure 4.2 shows two plots. On the left, we can see the OLS regression estimated on the sample data with no weights, and on the right, we display the OLS regression that has taken into consideration the RECS 2015 sampling weights. As it can be seen from the figure, the size of the dots on the right figure is proportional to the number of households

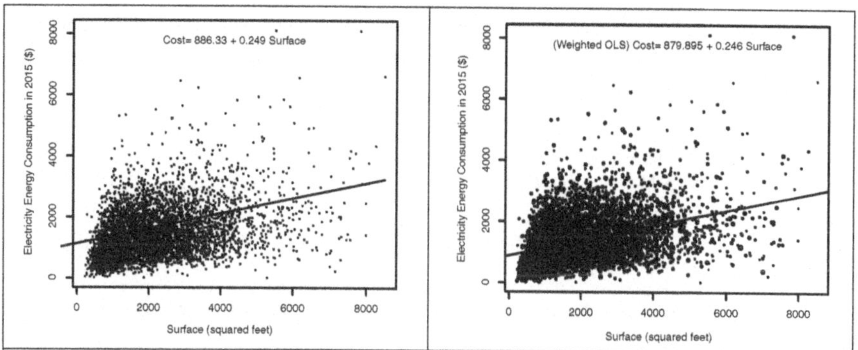

(Unweighted analysis)	(Weighted analysis)
Mean Cost= 1403.8 (s.d.=10.8)	Mean Cost= 1374.7 (s.d.=11.6)
Mean Surface= 2081.4 (s.d.=170)	Mean Surface= 2008.5 (s.d.=18.4)

Fig. 4.2 OLS regression of energy consumption on square footage unweighted (left) and using the sample weights (right). R program (see Appendix). *Source* Illustration by the author

that each observed household represents. As a result, it becomes clear that including the sampling weights in the implementation changes the descriptive results and the simple regression analysis.

In Fig. 4.2, the difference between the unweighted sample (left) and the weighted sample (right) analysis shows that there are slight differences in the estimation of the mean cost, the mean surface and the regression line when comparing these two cases. The results on the left would only describe the particular sample of households. Those on the right represent the universe of US households that are represented by the weighted sample of households provided by RECS 2015.

4.3 Quantile Regression in a Simple One-Covariate Model

We present two examples to analyze the weighted quantile regression applied to energy consumption. The first one aims to study how the total square footage of each household affects the distribution of energy consumption, and the second one aims to analyze how other independent variables detailed in Table 4.1 affect the distribution of energy consumption.

Table 4.2 shows the estimates by an OLS regression and by four additional quantiles regressions of $\theta = \{0.25, 0.50, 0.75, 0.95\}$ to identify possible changes of energy consumption along the quantile levels. We have only presented here the case, which corresponds to including the weights, which is the recommended approach in this example with data coming from a complex survey.

Table 4.2 Weighted OLS and weighted quantile regressions' coefficients at different quantiles

Cost (energy consumption)	OLS regression	$\theta_{0.25}$	$\theta_{0.50}$	$\theta_{0.75}$	$\theta_{0.95}$
Intercept	879.9* [843.8; 915.1]	464.45*,+ [427.03; 501.88]	749.73*,+ [704.03; 795.43]	1 132*,+ [1066.87; 1197.68]	1 789.97*,+ [1668.34; 1911.62]
Surface (total square footage)	0.25* [0.23; 0.26]	0.19*,+ [0.17; 0.21]	0.24* [0.22; 0.26]	0.31*,+ [0.28; 0.34]	0.48*,+ [0.40; 0.56]

[a; b] 95% confidence interval for each regression coefficient, a represents the lowest boundary and b the highest boundary
R program (see Appendix)
Source Calculations by the authors
*Significantly different from zero at the 5% significance level
+Significantly difference between quantile regression coefficients and the OLS coefficients at the 5% significance level

If effects along the variable distribution remain constant or similar, there is no point for using quantile regression because the variable distribution effects far from the mean might be quite similar to the effects obtained with OLS coefficients for the mean value.

In addition, two types of significance analysis are necessary to carry out in a quantile regression. We need to check whether or not quantile coefficients are significantly different from zero, and whether or not quantile coefficients are significantly different from OLS coefficients.

Now we need to see if quantile regression differs from the OLS approach. How should be these types of statistical significance be detected or computed? We propose one possible approach: The first one denoted as * in Table 4.2 is the traditional confidence interval of the estimate. For example, if it contains the zero, we do not reject that the corresponding coefficient is not significantly different from zero. When we look at the differences between linear regression and quantile regression, we need to look at the second symbol in Table 4.2 denoted as +, which appears if the confidence interval of the quantile regression coefficient does not overlap the OLS coefficient. If that happens, then both regression coefficients are accepted to be significantly different.

Table 4.2 shows that all the quantile regression coefficients except total square footage when $\theta = 0.50$ are statistically different from the OLS regression. This means the influence of the building surface affects the quantile of cost differently depending on the level of quantile that we are interested in.

Figure 4.3 shows that lower quantiles of the total square footage of a household consume less energy, while the highest ones tend to consume more energy. When the quantile is around 0.50, OLS regression confidence interval overlaps the quantile regression confidence interval, and therefore, we visually confirm that this specific quantile is not statistically different from the OLS regression coefficient. However, there is nothing to worry about, since we have already noted that there is a smooth

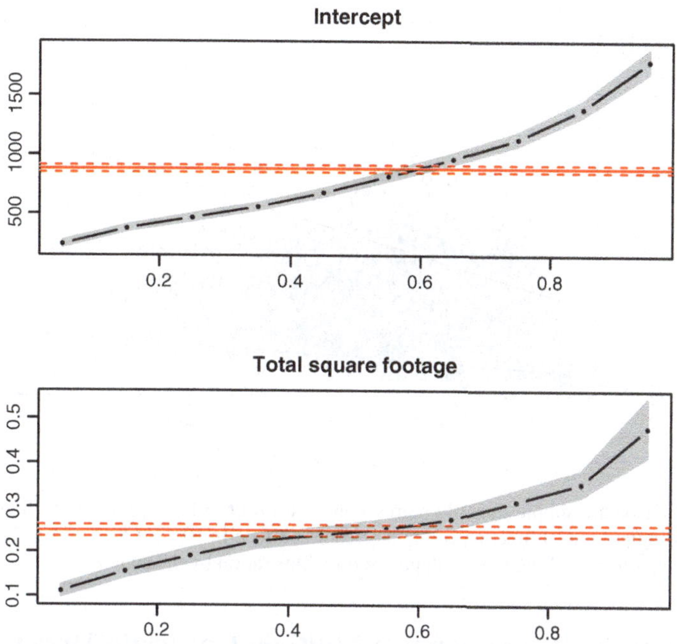

Fig. 4.3 Plot of the intercept (upper) and the coefficient of surface (lower) by quantile of the weighted quantile regression model and the OLS model in the energy consumption simple model for RECS (2015). *Note* The quantiles of the dependent variable are on the horizontal axis and the coefficient magnitudes on the vertical axis. The red line represents the OLS regression coefficient, and the red dotted lines represent the 95% confidence interval of the OLS regression coefficient. The dotted black line represents the quantile regression coefficient, and the gray boundaries represent the 95% confidence intervals of the respective quantile regression coefficient. R program (see Appendix). *Source* Illustration by the author

break in energy consumption after the 0.95 quantile. Thus, it is totally necessary to use quantile regression to focus on extreme energy consumers.

Therefore, running a quantile regression for the 0.95 quantile would be suitable for this case.

4.4 Coefficient Interpretation

If we take the results of the 0.95 quantile regression model, we would say that the effect of having one additional square footage increases the quantile of expenditure in 0.48 US$ in the total electricity cost, while the increase of one additional square foot impacts the lowest quantile of total electricity cost only around 0.19 US$ per square feet. In order to visualize the change of shape in the simple quantile regression, we plot all the quantiles in the sample plot. This is shown in Fig. 4.4.

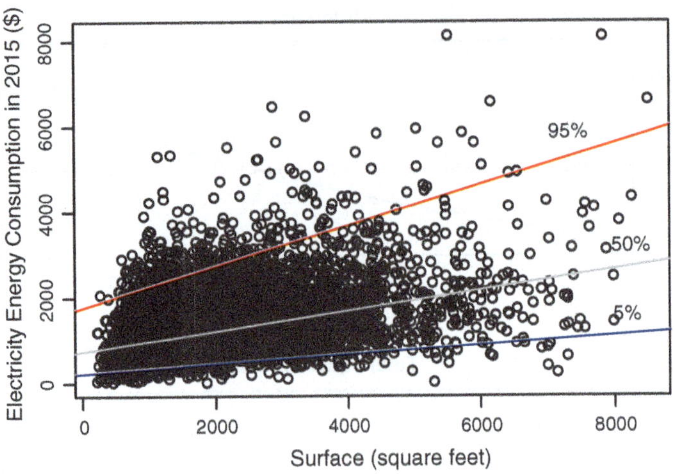

Fig. 4.4 Plot of the simple quantile regressions (5% lower, 50% medium, 95% higher) for the energy consumption as a function of the household surface in RECS (2015). *Note* The 5th, 50th and 95th quantiles are plotted in the figure. *Source* Illustration by the author

4.5 Quantile Regression in a Multiple-Covariate Model

In this second part of the chapter, we consider other determinants of electricity consumption. Previously, we have taken into account all covariates; however, we realize that most of them are highly correlated with each other, as for instance, the number of household members and the number of bedrooms or the number of smart phones. This is the reason why we will simplify our model to avoid multicollinearity issues. However, a quick look at Table 4.3 reveals that the number of members in the household and the existence of a swimming pool in the house are factors that affect the total consumption of electricity, but which effect changes depending on the considered quantile of expenditure.

Square footage, heated square footage and floor fans are not statistically different from zero according to Table 4.3. If we observe the behavior of these coefficients in Fig. 4.4, we easily realize that all the confidence intervals of the estimated coefficients contain zero along the conducted quantile regressions.

We can observe in Fig. 4.5 that, at 0.75 quantile, householders who have a swimming pool consume higher levels of energy with reference to householders who do not have a swimming pool. Being more specific, the 75% quantile of electricity cost increases by 770.58 US$ compared to those families that do not have a swimming pool. However, at this point, the 0.75 quantile regression coefficient associated with tenure of a swimming pool is not statistically different to the OLS regression parameter. At 0.95 quantile, owners with bigger swimming pools tend to have the highest levels of energy consumption and so the 95% quantile of electricity cost conditional

Table 4.3 Weighted quantile regressions' coefficients at different quantiles

Total energy consumption	OLS regression	$\theta_{0.25}$	$\theta_{0.50}$	$\theta_{0.75}$	$\theta_{0.95}$
Intercept	508.47* [428.14; 588.80]	288.39*,+ [243.86; 360.17]	471.93* [396.80; 545.04]	700.87* [562.74; 798.84]	1118.36*,+ [945.64; 1339.32]
Swimming pool	633.14* [561.50; 704.77]	487.96* [386.52; 561.56]	550.69* [463.27; 648,19]	770.58* [679.72; 934]	912.86*,+ [712.84; 1256.45]
Square footage	0.02 [−0.02; 0.05]	−0.01 [−0.05; 0.04]	0.03 [−0.01; 0.06]	0.00 [−0.04; 0.05]	0.10 [−0.01; 0.21]
Cooled square footage	0.12* [0.1; 0.15]	0.13* [0.10; 0.14]	0.14* [0.11; 0.16]	0.11* [0.08; 0.15]	0.09 [−0.02; 0.12]
Heated square footage	−0.02 [−0.06; 0.02]	−0.03 [−0.07; 0.02]	−0.08* [−0.11; −0.04]	−0.02 [−0.09; 0.02]	−0.00 [−0.09; 0.15]
Num. of bedrooms	82.29* [54.62; 109.98]	54.12* [36.20; 77.64]	91.84* [63.14; 117.17]	97.49* [53.35; 140.05]	139.19* [74.51; 184.55]
Electricity generation from solar	243.82* [84.32; 403.32]	285.29* [61.11; 369.75]	148.85* [38.70; 182.24]	170.45 [−110.72; 452.86]	1050.29 [−284.88; 3191.41]
Num. TVs	103.51* [86.29; 120.72]	91.10* [74.23; 102.40]	91.61* [70.16; 109.49]	128.83* [102.62; 155.85]	128.29* [62.44; 183.09]
Smart phones	21.62* [2.74; 40.49]	19* [3.54; 36.14]	36.08* [16.92; 48.32]	30.06* [0.02; 51.07]	−40.80 [−84.16; 7.81]
Floor fans	−3.79 [−21.97; 14.41]	−0.10 [−22.55; 13.88]	−7.86 [−24.51; 17.54]	2.43 [−24.49; 29.62]	11.73 [−29.88; 72.07]
Num. of household members	79.01* [61.66; 96.36]	59.59* [43.02; 74.74]	68.92* [59.30; 94.81]	107.13* [90.87; 138.79]	181.84*,+ [127.58; 250.1]

[a; b] 95% confidence interval for each regression coefficient, a represents the lower bound and b the higher boundary

*Significantly different from zero at the 5% significance level

+Significantly quantile regression coefficients from the OLS coefficients at the 5% significance level

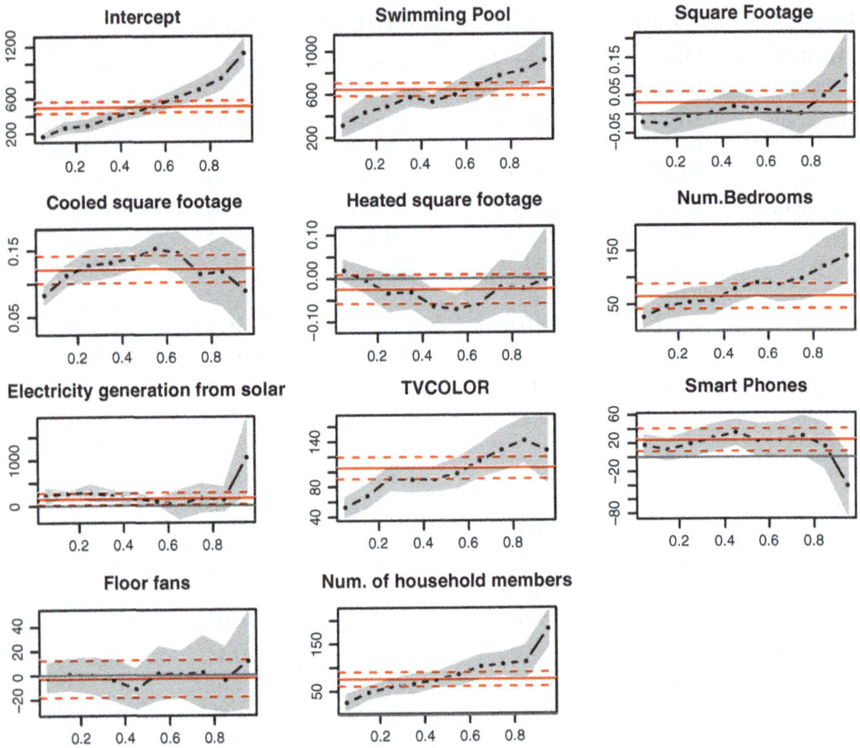

Fig. 4.5 Plot of the coefficients of the weighted quantile regression model and the OLS model for electricity consumption RECS (2015). *Note* The figure shows the effect of each of the explanatory variables on household energy consumption. The horizontal axis in each subplot refers to the estimation quantile, while the vertical axis is the level of the effect. Dotted black lines are quantile regression coefficients, and they are accompanied by their respective confidence intervals (shadowed area). Solid red lines in each subplot correspond to the OLS effect, and the associated confidence intervals are plotted as parallel dotted lines on each side of the OLS coefficient. *Source* Illustration by the author

on all other factors increases by 912.86 US$ for those having a swimming pool compared to those that do not have a swimming pool.

If we see cooled square footage, number of bedrooms and TVCOLOR in Fig. 4.4, their quantile coefficients are not statistically different from the OLS coefficients; both confidence intervals overlap through the quantile regressions for the moderate quantile levels. Thus, in order to study the relationship between energy consumption and the cooled square footage of a household, an OLS regression can be used. At 0.95 quantile, one additional household member increases the 95th quantile of the total electricity bill by 181.84 US$ in total electricity bills.

4.6 Conditional Versus Unconditional Quantile Regression

The most frequent use of quantile regression is conditional quantile regression, because the interest is on understanding the impact of covariates on the distribution of a response. However, the difference between conditional and unconditional quantile regression is on the focus. Conditional quantile regression focuses on the quantile of the output for fixed values of the covariates. So, given the covariates, conditional quantile regression predicts the θth quantile of the outcome.

In contrast, unconditional regression quantiles are interested in the marginal distribution of the response, regardless of the values of the covariates. It can be shown that the partial effect of a covariate on an unconditional quantile of dependent variable Y is expressed as a weighted average of the partial effect on a specific conditional quantile of Y that corresponds to the unconditional quantile that is of interest (Firpo et al. 2009). In fact, the 95% percentile of the unconditional distribution of Y may not be the same as the 95% percentile of the conditional distribution of Y for fixed values of the covariates.

The examples given in the previous sections have served to understand the influence of covariates on the quantile of consumption cost. For instance, in the single-covariate model, we have concluded that one additional square foot in the building surface increases the 95% percentile of consumption cost by 0.48 dollars (see Table 4.2). The interpretation is linear, so 95% households are predicted to pay each year a cost for electricity consumption below the value that is predicted by the 95% quantile regression estimate, given their corresponding surface.

Unconditional quantile regression is a different model. In that case, we would be interested in knowing the impact of an increase of one squared foot in the average surface countrywise on the 95% quantile of variable energy cost. So, if households increased the building's surface on average, we would like to see the impact on the quantile of the energy cost distribution.

We have implemented the estimation of unconditional quantile regression in our context. See Table 4.4.

Table 4.4 Weighted OLS and unweighted unconditional quantile regressions' coefficients at different quantiles

Cost (energy consumption)	OLS regression	$\theta_{0.25}$	$\theta_{0.50}$	$\theta_{0.75}$	$\theta_{0.95}$
Intercept	879.9* [843.8; 915.1]	467.73* [449.03; 486.44]	793.88* [770.15; 817.6]	1239.54* [1198.52; 1280.56]	1904.73* [1752.96; 2056.51]
Surface (total square footage)	0.25* [0.23; 0.26]	0.17* [0.16; 0.18]	0.22* [0.21; 0.23]	0.28* [0.26; 0.30]	0.47* [0.39; 0.54]

*Significantly different from zero at the 5% significance level. Unconditional quantile regression was estimated without sampling weights

What we learned from implementing unconditional quantile regression is that the impact of an increase in the average size of apartments on the quantiles of the energy cost distribution is significant. So, an additional square foot in average building surface would increase the 95% percentile of the cost distribution in our group of sampled households by 0.47 dollars.

4.7 Summarizing Remarks

Quantile regression aims to study the effects of independent variables on the conditional quantile of the dependent variable, while the OLS regression studies the effect of the independent variables on the conditional mean of the dependent variable. Therefore, before running model regression, it is worth value plotting the quantiles of the dependent variable to detect a possible pattern in its distribution at least unconditionally.

We also recommend testing individual significances so as to prove that is necessary to run a quantile regression rather than just an OLS. However, in order to run a quantile regression, heteroskedasticity must be detected.

Finding the appropriate quantile for running a quantile regression requires finding a break or change through the quantile coefficients.

References

Firpo, S., Fortin, N. M., & Lemieux, T. (2009). Unconditional quantile regressions. *Econometrica, 77*(3), 953–973.

Pfeffermann, D., Skinner, C. J., Holmes, D. J., Goldstein, H., & Rasbash, J. (1998). Weighting for unequal selection probabilities in multilevel models. *Journal of the Royal Statistical Society: Series B (Statistical Methodology), 60*(1), 23–40.

Solon, G., Haider, S. J., & Wooldridge, J. M. (2015). What are we weighting for? *Journal of Human Resources, 50*(2), 301–316.

Chapter 5
Time Series Quantile Regression

Abstract In this chapter, we explore a topic that has gained considerable attention in the academic literature during the latter years, namely quantile regression for time series data. We will illustrate how to use conditional quantile regression to model the dynamics of electricity prices as a function of the price of an input and also substitute energy commodity such as natural gas. We will also illustrate how to estimate different conditional quantiles of electricity prices using lags of the same prices, in the vein of traditional autoregressive processes in time series analysis. This exercise when conducted to estimate the tail quantiles (i.e., very low or very high quantiles) allows us to recover a modified value at risk statistic on the price distribution, which is asymmetric conditional on each tail, and which can be used to monitor the market risk for consumers and producers of electricity.

Keywords Time series · Energy prices · Autoregressive process · Price dynamics · Energy operator · Unit root · Extremes · Market risk

We will present in what follows two examples that use real price data for the Pennsylvania-New Jersey-Maryland (PJM) interconnected hub of the US electricity market. We have chosen the PJM interconnection for our time series analysis of electricity and natural gas prices, given its importance for the US wholesale electricity market.

It is widely known that the price of natural gas is an important determinant of electricity prices, because it is an input for electricity generation (for instance, natural gas represents 24% of PJM's electricity generation mix); but also because natural gas and electricity are substitutes for heating in households and firms. Thus, it may be expected for natural gas prices to affect electricity prices in a significant fashion. For instance, an increment in the demand for natural gas triggered, by abnormally low temperatures, will translate into an increment of natural gas prices given a relatively inelastic supply. These higher natural gas prices will be associated with higher generation costs for natural gas-fired power generators, and such costs will be passed through final electricity consumers.

Supply and demand dynamics in electricity markets are featured by prominent seasonality, which is naturally reflected in nonlinear price dynamics of both electricity inputs and outputs. This fact motivates our approach, which considers nonlinearities

J. M. Uribe and M. Guillen, *Quantile Regression for Cross-Sectional and Time Series Data*, SpringerBriefs in Finance, https://doi.org/10.1007/978-3-030-44504-1_5

across the conditional quantiles of the energy price distribution. The use of quantiles allows us to directly model electricity prices seasonality and more importantly to access the possible nonlinear effects that describe the relationship between natural gas prices and electricity prices. In this way, by focusing on the quantiles of the price distribution instead of the conditional mean, we are able to measure the impact of natural gas on electricity when the latter is relatively cheap (low quantiles) or relatively expensive (high quantiles). This sort of analysis increases our understanding of the relationship between the two energy commodities. In the same way, we also are able to construct a more accurate representation of the time series structure that better describes electricity prices and their seasonality by allowing nonlinear impacts of past electricity prices on the price today. See Uribe et al. (2018) for a close examination that unlike here uses a quantile-quantile approach.

5.1 Data Source

The largest independent system operator (ISO) in the US, a fully restructured competitive market, which has a geographical footprint of 243,417 square miles and an installed capacity of 165 GW, is the Pennsylvania-New Jersey-Maryland interconnection. This ISO covers 13 states (Delaware, Illinois, Indiana, Kentucky, Maryland, Michigan, New Jersey, North Carolina, Ohio, Pennsylvania, Tennessee, Virginia, West Virginia and the District of Columbia) with a population of 65 million. It was founded in 1927 with a pool of three utilities New Jersey and Pennsylvania. In 1956, Maryland joined the interconnected mechanism, and by 1996 PJM was already a totally functioning ISO. We use the daily spot electricity and natural gas prices for the PJM hub as published by the US Energy Information Administration (EIA). Prices are daily, and our sample starts in March 2014 and ends in December 2017.

Table 5.1 reports summary statistics and unit root test (augmented Dickey–Fuller—ADF) of the price data. PJM WH Real Time Peak is the ICE electricity product name for the PJM West Hub, and TETCO-M3 is the name given to the product originating from the natural gas hub TETCO-M3, also in the PJM region. Regarding the ADF test, there is evidence against the presence of unit root in the series of prices. This is important as it allows us to conduct the quantile regressions in levels, instead of having to resort to transformations of the original series. As in ordinary regressions, variables in levels that contain a unit root could be used only if their joint behavior is stationary. That is if the set of variables are cointegrated. Otherwise, the variables should be transformed before estimation as to ensure a correct interpretation and estimation of the quantile slopes and other parameters in the model.

For further exploration of electricity prices, which will become our dependent variable in the following estimations, we present Fig. 5.1. It shows the electricity price's quantiles plotted against the corresponding fraction of the data, sorted from the lowest to the highest values. With this graphical representation, we seek to characterize how the variable's quantiles behave. Thus, we observe that there is a low price

Table 5.1 Summary statistics and ADF test for wholesale electricity and natural gas spot prices in the PJM Market

	PJM ($/MWh)	TETCO-M3 ($/MMBtu)
Mean	39.34	2.39
Std. dev.	15.38	1.86
Minimum	22.70	0.38
25th percentile	31.15	1.43
50th percentile	36.03	1.91
75th percentile	42.98	2.85
Maximum	231.50	20.65
Observations	941	941
t-ADF	−5.87[a]	−4.88[a]

$/MWh stands for dollar per megawatt hour and $/MMBtu stands for dollar per million British thermal units

[a]The null hypothesis of a unit root is rejected at 1% level of confidence. Sample period March 2014–December 2017

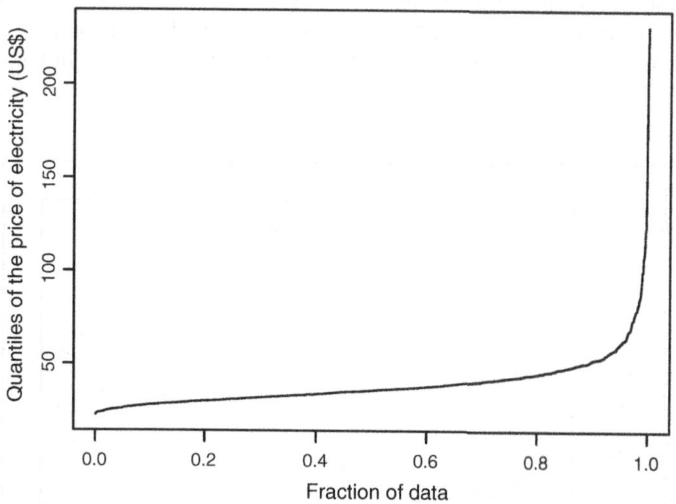

Fig. 5.1 Quantiles for the price of electricity (US$) EIA. R program, in the Appendix. *Source* Illustration by the author

of electricity for the lower quantiles, and around 28 US$ for the median price with a smooth increase until the 90% quantile approximately. But after that, its distribution has a sharp peak, which interestingly is not present in the lower quantiles.

5.2 Natural Gas Prices as a Determinant of Electricity Prices—An OLS Example

We start our analysis of the relationship between electricity and natural gas prices by regressing the first variable on the latter. This can be seen as a "mean" to "mean" approach, in which we consider the effects of an average change (unitary change) in natural gas prices on the average change (the conditional mean) of electricity prices. As can be seen in Fig. 5.2, in which this relationship is plotted, a unitary increment (1 $/MMBtu) of the natural gas price is associated with an increase of about 6 $/MWh in the price of electricity. This relationship is statistically significant at 99% level of confidence (s.e. 0.186) and it confirms the theoretical relationship that emerges because natural gas simultaneously behaves as an input and a substitutive of electricity. It is also possible to note from the same figure that some outliers in both series (simultaneous abnormally high prices of the two commodities) may be asserting some influence on the mean results reported for the OLS regression.

Situations, in which both natural gas and electricity prices are abnormally high, are naturally related to episodes of relatively low energy supply, either because of an abnormal increase in the demand for electricity (for instance, after a wage of cold weather that fosters heating energy consumption by households), or because of a reduction in the energy supply (following, for instance, curtails in the electricity generation by some variable renewable energy sources in the energy mix of the

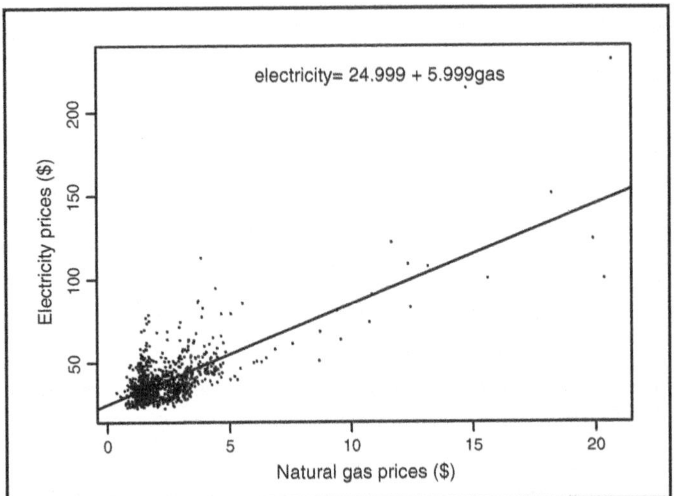

MeanElectricity = 3 9.335(s.d. = 1 5.375)
MeanNaturalgas = 2.389(s.d. = 1.857)

Fig. 5.2 OLS regression of electricity prices natural gas prices. R program (see Appendix). *Source* Illustration by the author

market). These situations may be directly addressed by the means of conditional quantile regressions, in which the analysis allows for asymmetries in the relationship between electricity and natural gas prices, emerging at different fragments of the conditional distribution of electricity prices. This, in turn, will allow us to consider seasonality in energy prices that otherwise cannot be directly modeled.

5.3 Quantile Regression in a Simple One-Covariate Model

In what follows, we extend the linear regression analysis conducted before as to incorporate possible changes in the relation between electricity and natural gas prices by means of conditional quantile regressions. Analyzing conditional quantile effects instead of conditional mean effects is useful to study the impact of changes in the price of natural gas on electricity when electricity is relatively cheap or relatively expensive. It also allows us to study scenarios of possible materialization of energy crisis (for instance, due to a simultaneously insufficient provision of the two commodities).

Table 5.2 presents the estimates by an OLS regression and for nine regression quantiles $\theta = \{0.1, 0.2, 0.3, 0.4, 0.50, 0.6, 0.7, 0.8, 0.9\}$. Recall that whether the effects along the variable distribution remain constant or similar, there is little room for improvement in the analysis fitting quantile regressions to the data, because in this case, the variable distribution is around the mean, and therefore OLS coefficients suffice to describe the relationship between the two energy prices. This fact motivates two types of significance analysis, which are to be conducted when using conditional quantile regressions: First, we need to verify that the quantile slope coefficients are significantly different from zero, and second that the quantile coefficients are significantly different from OLS coefficients.

There are different ways to calculate the statistical differences between the quantiles slopes and zero on the one hand and the quantile and OLS slopes in the other hand. One direct way to estimate them consists of noticing whether the zero (or the OLS coefficient) lies within the interval associated with each quantile slope. When this is the case the statistical test does not render a significant difference and the quantiles slopes are not statistically different from zero (former case), or they are not different than the OLS slopes (in which case the regression quantiles do not offer deeper insights for the problem at hand on top of the traditional OLS regression). The results regarding the first statistical test are denoted by * and they might be confirmed using the associated confidence interval of the estimate. The second test is denoted as + and it might be recorded if the confidence interval of the quantile regression coefficient does not overlap the OLS coefficient.

Table 5.2 shows that all the quantile regression slopes except natural gas prices when $\boldsymbol{\theta} = 0.7$ and $\boldsymbol{\theta} = 0.8$ are statistically different from the OLS regression. In the same vein, the intercepts of the quantile lines are also statistically different from the OLS intercept except for $\boldsymbol{\theta} = 0.4$.

Figure 5.3 depicts a clear pattern of the relationship between electricity and natural gas prices. That is, the impact of changes of natural gas prices on electricity

Table 5.2 OLS and quantile regression coefficients at different quantiles

Electricity price	OLS	$\theta_{0.1}$	$\theta_{0.2}$	$\theta_{0.3}$	$\theta_{0.4}$	$\theta_{0.5}$	$\theta_{0.6}$	$\theta_{0.7}$	$\theta_{0.8}$	$\theta_{0.9}$
Intercept	25.0*	20.5*,+ [19.6; 22.3]	22.7*,+ [21.0; 23.4]	23.5*,+ [22.2; 25.1]	25.0* [23.9; 26.6]	26.4*,+ [25.8; 27.4]	27.7*,+ [19.6; 22.3]	28.6*,+ [27.2; 30.5]	29.1*,+ [28.5; 31.2]	32.7*,+ [27.9; 36.7]
Natural gas price	6.0*	3.67*,+ [2.47; 4.04]	3.82*,+ [3.26; 4.67]	4.29*,+ [3.72; 4.82]	4.56*,+ [3.74; 4.88]	4.61*,+ [4.20; 4.92]	4.77*,+ [4.37; 5.44]	5.54* [4.60; 6.15]	6.63* [5.54; 7.08]	7.66*,+ [6.35; 9.79]

Source Calculations by the authors

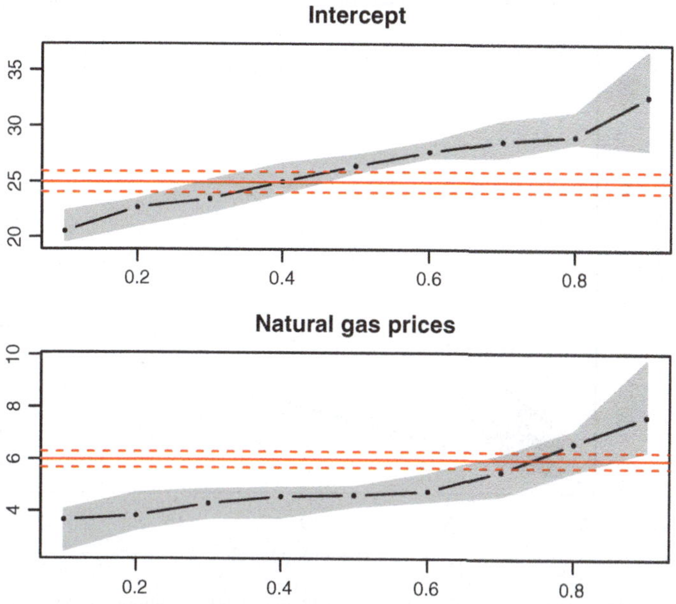

Fig. 5.3 Plot of the intercept (upper) and the coefficient of natural gas (lower) by quantile in the model of electricity prices. *Note* The quantiles of the dependent variable are on the horizontal axis and the coefficient magnitudes on the vertical axis. The red line represents the OLS regression coefficient, and the red dotted lines represent the 95% confidence interval of the OLS regression coefficient. The choked black line represents the quantile regression coefficient, and the gray boundaries represent the 95% confidence interval the quantile regression coefficient. R program (see Appendix). *Source* Illustration by the author

prices is greater the higher the prices and quantile of electricity. The OLS and quantile slopes only overlap when the price of electricity is between its 70th and 80th percentiles. Otherwise, they OLS slopes and the quantile slopes that measure the impact are statistically different from each other, which confirm the convenience of using quantile regression to fully understand the way in which these two energy commodities interact.

5.4 Coefficient Interpretation

One interesting feature of quantile regressions is that they allow us to compare how increments in natural gas prices are translated into increments in electricity prices, when electricity is relatively expensive (high quantiles) or relatively cheap (low quantiles). For instance: an increment of 1 \$/MMBtu in the price of natural gas is related to, or even causes, an increment of 3.67 \$/MWh in the price of electricity, when the price of the latter is at $\boldsymbol{\theta} = 0.1$, that is, when electricity is relatively cheap.

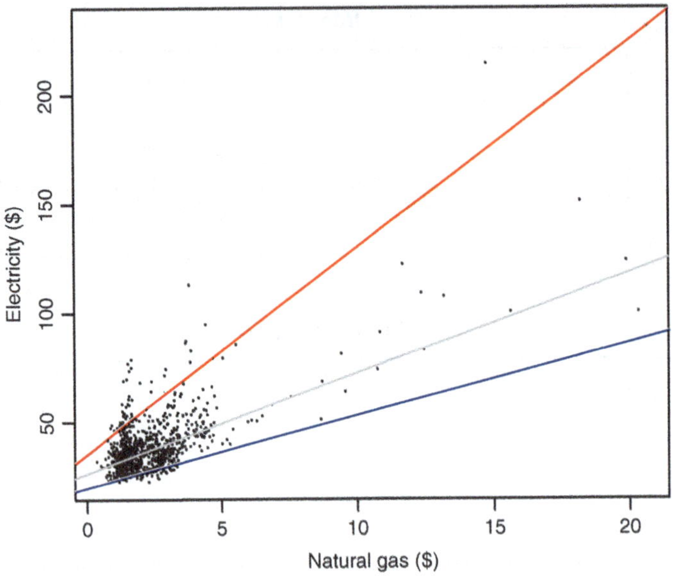

Fig. 5.4 Plot of the simple quantile regressions (5% lower, 50% medium and 95% higher) for the electricity prices as a function of the natural gas prices. *Note* The 5th, 50th and 95th quantiles are plotted in the figure. *Source* Illustration by the author

On the other hand, an increment of 1 \$/MMBtu in the price of natural gas is associated with an increment of 7.66 \$/MWh in the price of electricity, when the price of the latter energy commodity is at $\theta = 0.9$, in other words, when electricity is relatively expensive. See a graphical representation of this in Fig. 5.4. All in all, the impact of natural gas on electricity prices intensifies as electricity becomes relatively scarce and therefore more expensive.

5.5 Autoregressive Quantiles

In the second part of this chapter, we consider models in the vein of traditional autoregressive specifications. That is, we use lags of the price of electricity in order to forecast different quantiles of the electricity price distribution. By so doing, we are able to calculate a statistic similar to the widely used value at risk (VaR) statistic in finance, which consists of a very high quantile of the loss distribution of a given portfolio. The VaR measures the maximum loss that an investor may expect to face over a certain horizon, usually a day or a week, with a certain level of confidence, usually 95%, 99% or 99.9%. Provided the VaR is a quantile, such a level of confidence naturally corresponds to a percentile of the loss distribution. The higher the quantile, the lower the probability that an observed loss realization overpasses the quantile value during the holding period. In our case, the VaR quantile is calculated using the price distribution instead of the loss function, and therefore it cannot be interpreted as a potential loss, but as an unusually high or low electricity price. Nevertheless, our VaR-alike statistic will keep its utility for risk analysis for consumers and producers of electricity. High quantiles are in order to estimate the risk of consumers of electricity, while low quantiles will put at risk future revenues of producers.

In Table 5.3, we present the AIC criteria used to estimate the number of lags when modeling different quantiles of electricity prices. As in traditional time series analysis, AIC must be minimized in order to select the better model specification, which is the one with the better balance between a high explanatory power and less parameters to be estimated (in order to preserve the degrees of freedom of the estimation and minimize the risk of overfitting). As may be observed from the table, different quantiles can be explained by a different number of lags of the electricity prices. This occurs because, as explained in Chap. 3, the quantile penalty function is asymmetric, and it changes according to the estimated quantile, assigning more or less weight to different realizations in the minimization problem, according to how far a given realization is from a given quantile of interest.

The AIC formula is given by:

$$AIC = 2k + 2N \log RSS,$$

where k is the number of parameters in the model, N the sample size and RSS stands for the residual sum of squares.

In our case, we restrict the minimum number of lags to 1 and the maximum to 5. Quantiles $\boldsymbol{\theta} = 0.1, 0.3, 0.6, 0.8$ are better described by the maximum lags, while quantiles $\boldsymbol{\theta} = 0.2, 0.4$ are by the minimum. And for $\boldsymbol{\theta} = 0.5, 0.7, 0.9$ the optimum number of lags is 4, 3, 2, respectively.

In Fig. 5.5, we can observe the coefficients associated with different lags of the price of electricity across different quantiles of the electricity price distribution. We can see that the persistence of an impact past electricity price variations on the current quantiles of the electricity prices varies greatly for the first lag, ranging from 0.63 to 1.28. For the other lags, such nonlinearity across the quantiles can also be detected, but is not that strong.

In Fig. 5.6, we plot electricity prices and percentiles 10th (left) and 90th (right), following a model that included five lags of prices as explanatory variables. As can be seen, the conditional quantiles change in time, according to the value of the covariates. This allows to forecast one-day ahead quantiles of electricity prices, and thus constructing risk scenarios for consumers and producers of electricity. Notice that the low quantiles such as the 10th percentile are particularly important for producers of electricity that commit to supply a given amount of power to the market at a certain price. On the other hand, a high quantile, such as the percentile 90th are particularly important for consumers that must cover their energy demand at a given market price.

Table 5.3 AIC criteria for quantile models with autoregressions

Lags	$\theta_{0.1}$	$\theta_{0.2}$	$\theta_{0.3}$	$\theta_{0.4}$	$\theta_{0.5}$	$\theta_{0.6}$	$\theta_{0.7}$	$\theta_{0.8}$	$\theta_{0.9}$
1	22,566.92	21,503.99*	22,160.20	21,615.67*	21,847.22	21,966.00	21,633.50	23,124.78	21,484.34
2	22,135.77	21,578.27	21,857.39	22,045.17	21,645.84	22,964.41	21,536.26	22,470.00	21,457.58*
3	21,860.64	22,067.23	21,675.26	22,989.15	21,543.39	22,508.54	21,504.30*	22,048.86	21,538.22
4	21,664.73	22,996.23	21,538.88	22,502.47	21,512.04*	22,065.46	21,600.59	21,815.32	22,036.15
5	21,522.19*	22,565.99	21,511.20*	22,077.30	21,591.60	21,854.63*	22,049.32	21,606.16*	23,063.52

Source Calculations from the author

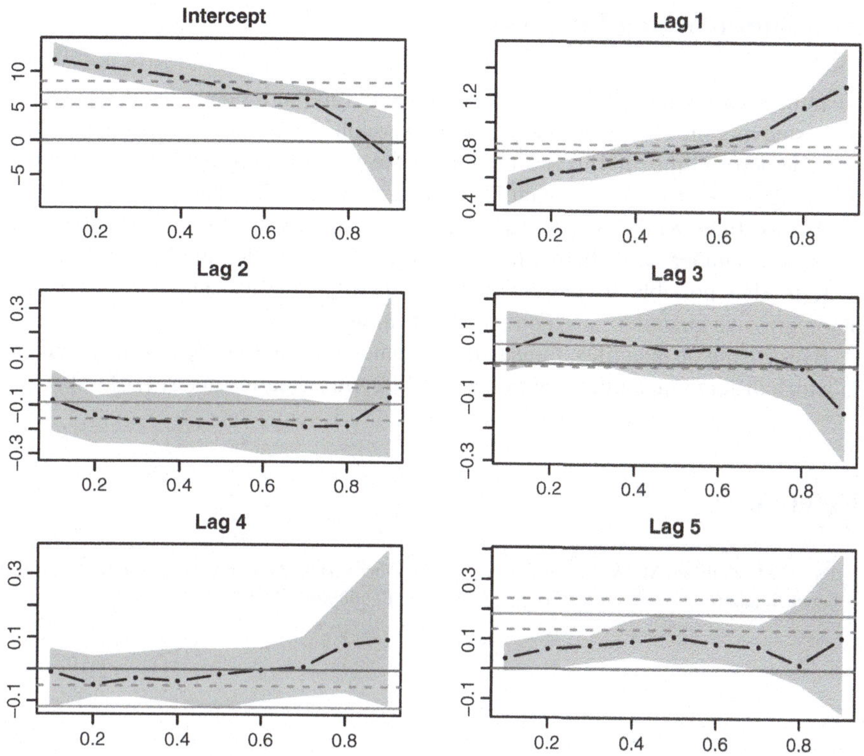

Fig. 5.5 Plot of the intercept and the coefficients associated to different lags of the price of electricity at different quantiles. *Note* The quantiles of the dependent variable are on the horizontal axis and the coefficient magnitudes on the vertical axis. The red line represents the OLS regression coefficient, and the red dotted lines represent the 95% confidence interval of the OLS regression coefficient. The dotted black line represents the quantile regression coefficient and the gray boundaries represent the 95% confidence intervals of the respective quantile regression coefficient. R program (see Appendix). *Source* Illustration by the author

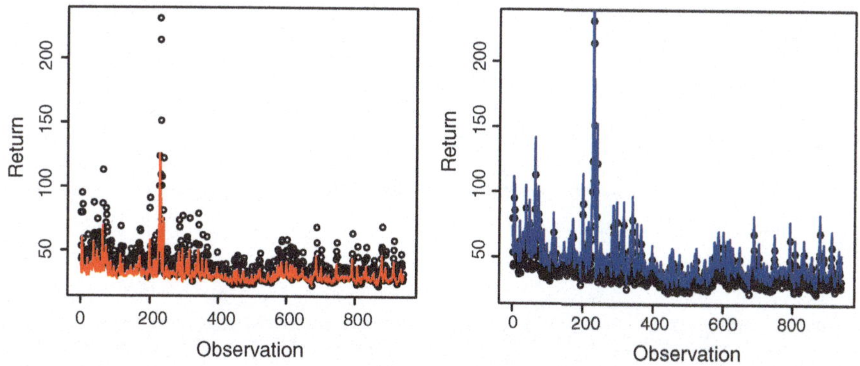

Fig. 5.6 Plot of electricity prices and percentiles 10th (left) and 90th (right), including five lags of prices as explanatory variables. *Source* Illustration by the author

5.6 Summarizing Remarks

Quantile regression is a useful tool for analyzing time series data. It can be used for both, studying the effects of an explanatory variable on the quantiles of an explained variable across time, and to run models in the vein of traditional time series data using lags to forecast future quantiles of the conditional distribution.

When estimating the time series quantiles is necessary first to verify the stationary of the series under study, before fitting the quantile regression model to the data.

It is also possible to employ traditional AIC or BIC statistics to estimate autoregressive quantiles based on past realizations.

By focusing on quantiles, forecasting becomes a useful tool for risk management, in which different quantiles correspond to different scenarios in terms of risk.

Reference

Uribe, J. M., Guillen, M., & Mosquera, S. (2018). Uncovering the nonlinear predictive causality between natural gas and electricity prices. *Energy Economics, 74*, 904–916.

Chapter 6
Goodness of Fit in Quantile Regression Models

Abstract In this chapter, we discuss the goodness of fit in the context of quantile regression. We present the coefficient of determination and its adaptation to the quantile regression context. We illustrate how to calculate this goodness of fit statistic and how to interpret it using data for energy consumption, its determinants and electricity prices.

Keywords Goodness of fit · R squared · Model evaluation · Conditional quantile · AR model · Model comparison

One popular statistic, used in the context of traditional modeling of the conditional "mean" effects of a given variable, is the coefficient of determination. This coefficient of determination, R^2, is defined in the following way in the linear regression framework:

$$R^2 = 1 - \widehat{S}/\widetilde{S},$$

where $\widetilde{S}$ refers to the sum of squares of a restricted model, usually restricted as to include only an intercept within the set of regressors. In other words, $\widetilde{S}$ is the total sum of squares of the explained variable, calculated around its unconditional mean. On its side, $\widehat{S}$ corresponds to the sum of squares of the unrestricted model, or in other words, it stands for the residual variation left unexplained by a given linear model.

By construction $\widehat{S} < \widetilde{S}$ and thus R^2 lies between zero and one. Zero means that the model at hand does not explain *at all* the variation of the predicted variable on the left-hand side of the equation, and one means that the studied model explains *the whole variation* of the predicted variable. Naturally, in real applications, R^2 is never zero or one, but it lies in-between. The larger the R^2 value, the greater the fitting of actual the model to the data.

J. M. Uribe and M. Guillen, *Quantile Regression for Cross-Sectional and Time Series Data*, SpringerBriefs in Finance, https://doi.org/10.1007/978-3-030-44504-1_6

Koenker and Machado (1999) propose a natural generalization of the coefficient of determination to the case of conditional quantile regression, given by the following formula:

$$R^2(\theta) = 1 - \widehat{V}(\theta)/\widetilde{V}(\theta),$$

where analogously to the linear counterpart explained before, $\widetilde{V}(\theta)$ stands for the square residuals of the restricted minimization problem specific for the case of the θ quantile, and $\widehat{V}(\theta)$ for the unrestricted statistic. Once again, $R^2(\theta)$ lies between zero and one. Unlike R^2 which measures the relative success of two models (restricted and unrestricted) for the *conditional mean function* in terms of residual variance, $R^2(\theta)$ is a measure of the relative success of the corresponding quantile regression model at a specific quantile, this time in terms of a weighted sum of absolute residuals. Therefore, $R^2(\theta)$ can be seen as a local measure of goodness of fit (for a particular quantile), rather than a global measure of the goodness of fit, which holds (in average) across the whole conditional distribution of the predicted variable, as is the case for R^2.

In what follows, we calculate the $R^2(\theta)$ for the case of an AR(1) model at different quantiles of the response variable, following the example developed across this chapter. That is, we will explore how much of the variation of electricity prices is explained by the first lag of electricity prices at different quantiles of the variable.

In Table 6.1, we show the coefficient of determination for an AR(1) model of the prices of electricity, estimated at three different quantiles of the conditional distribution of electricity prices, namely at the 10th, 50th and 90th percentiles. As can be observed, the variation explained by the first lag of the prices ranges from a minimum of 49.99% for the 90th percentile, to a maximum of 60.23% at the median of the variable.

The $R^2(\theta)$ might help to complement the analysis presented in Table 6.1 of this chapter, as it is a measure of fit to the data, and can be used to compare different models with the same predicted variable. Nevertheless, its appealing emerges because unlike the AIC statistic it provides an interpretable measure of the goodness of fit, which helps to construct an order of magnitude of the model success when it is fitted to a specific data set.

In the cross-sectional analysis of energy household expenditure, the calculation of the goodness of fit produces the results that are lower than the OLS coefficient of determination (R^2) in the simple case of one covariate (see Table 6.2).

When we look at models with a larger number of covariates, then the conclusions indicate a substantial increase in the value of the statistic. We recommend not to

Table 6.1 Coefficient of determination of an AR(1) quantile regression model of electricity prices

$\theta_{0.1}$ (%)	$\theta_{0.5}$ (%)	$\theta_{0.9}$ (%)
52.93	60.23	49.99

Source Calculations from the author

Table 6.2 Goodness of fit for the weighted OLS and weighted quantile regressions' coefficients at different quantiles in the RECS data when energy consumption is explained by an intercept and the total surface of the house

Cost (energy consumption)	OLS regression	$\theta_{0.25}$	$\theta_{0.50}$	$\theta_{0.75}$	$\theta_{0.95}$
R^2 (%)	15.1	7.40	8.31	8.33	11.70

Source Calculations from the author

compare the coefficient of determination of the OLS estimation with the goodness of fit statistic resulting from the proposal by Koenker and Machado (1999). In general, we observe that the goodness of fit value increases when the quantile level is larger. This may be due to the fact that the information content of covariates if relatively more powerful than a null model with just the intercept in the right tail of the conditional distribution of the dependent variable. Here, energy consumption is rather right skewed, so the covariates seem to help to understand the reasons why some households tend to have a larger than median consumption, while they are less powerful helping to understand patterns associated with low-level consumption (see Table 6.3).

Table 6.3 Weighted quantile regressions' coefficients at different quantiles for the final models

Total energy consumption	OLS regression	$\theta_{0.25}$	$\theta_{0.50}$	$\theta_{0.75}$	$\theta_{0.95}$
Intercept	508.47* [428.14; 588.80]	288.39*,+ [243.86; 360.17]	471.93* [396.80; 545.04]	700.87* [562.74; 798.84]	1118.36*,+ [945.64; 1339.32]
Swimming pool	633.14* [561.50; 704.77]	487.96* [386.52; 561.56]	550.69* [463.27; 648.19]	770.58* [679.72; 934]	912.86*,+ [712.84; 1256.45]
Square footage	0.02 [−0.02; 0.05]	−0.01 [−0.05; 0.04]	0.03 [−0.01; 0.06]	0.00 [−0.04; 0.05]	0.10 [−0.01; 0.21]
Cooled square footage	0.12* [0.1; 0.15]	0.13* [0.10; 0.14]	0.14* [0.11; 0.16]	0.11* [0.08; 0.15]	0.09 [−0.02; 0.12]
Heated square footage	−0.02 [−0.06; 0.02]	−0.03 [−0.07; 0.02]	−0.08* [−0.11; −0.04]	−0.02 [−0.09; 0.02]	−0.00 [−0.09; 0.15]
No. of bedrooms	82.29* [54.62; 109.98]	54.12* [36.20; 77.64]	91.84* [63.14; 117.17]	97.49* [53.35; 140.05]	139.19* [74.51; 184.55]
Electricity generation from solar	243.82* [84.32; 403.32]	285.29* [61.11; 369.75]	148.85* [38.70; 182.24]	170.45 [−110.72; 452.86]	1050.29 [−284.88; 3191.41]

(continued)

Table 6.3 (continued)

Total energy consumption	OLS regression	$\theta_{0.25}$	$\theta_{0.50}$	$\theta_{0.75}$	$\theta_{0.95}$
No. of TV	103.51* [86.29; 120.72]	91.10* [74.23; 102.40]	91.61* [70.16; 109.49]	128.83* [102.62; 155.85]	128.29* [62.44; 183.09]
Smartphones	21.62* [2.74; 40.49]	19* [3.54; 36.14]	36.08* [16.92; 48.32]	30.06* [0.02; 51.07]	−40.80 [−84.16; 7.81]
Floor fans	−3.79 [−21.97; 14.41]	−0.10 [−22.55; 13.88]	−7.86 [−24.51; 17.54]	2.43 [−24.49; 29.62]	11.73 [−29.88; 72.07]
No. of household members	79.01* [61.66; 96.36]	59.59* [43.02; 74.74]	68.92* [59.30; 94.81]	107.13* [90.87; 138.79]	181.84*,+ [127.58; 250.1]
R^2 (%)	26.5	10.53	14.12	16.35	21.03

[a; b] 95% confidence interval for each regression coefficient, a represents the lower bound and b the higher boundary

Source Calculations from the author

*Significantly different from zero at the 5% significance level

+Significantly quantile regression coefficients from the OLS coefficients at the 5% significance level

Reference

Koenker, R. W., & Machado, J. A. F. (1999). Goodness of fit and related inference processes for quantile regression. *Journal of the American Statistical Association, 94*(448), 1296–1310.

Chapter 7
Novel Approaches in Quantile Regression

Abstract In this chapter, we describe three recent lines of research in quantile regression, the first one is rather an extension of nonparametric regression into the quantile regression framework. The second topics are an extension for time series quantile regressions of the traditional time series tool, the cross-correlogram. The third possible extension presents a random forest as an extension of traditional quantile regression.

Keywords Nonparametric methods · Kernel · Bootstrap · Cross-quantilogram · Q-test · Directional predictability · Random forests

7.1 Nonparametric Quantile Regression

One popular extension of the traditional quantile regression framework studied in the previous chapters is the nonparametric quantile regression (see for instance Cai 2002). The key difference between the two approaches lies on the fact that, while in a traditional quantile regression context, the aim is to estimate a *parameter (or set of parameters)* that describes the *linear function* that relates a change of the explanatory variable with an associated change of the conditional quantile in the response variable, in the nonparametric approach, the association between the explanatory variables and the various quantiles of the dependent variable is not restricted to be linear, instead it is only required to be a smooth and continuous function. In other words, in the nonparametric framework, the objective is to identify the best regression function according to the data distribution, rather than estimating parameters of a predetermined-fixed linear function.

There are three main nonparametric regression methods that have been adapted to estimate quantile-relationships between two or more variables. These methods are locally polynomial (kernel) methods, series methods and penalty methods.

Local polynomial methods assume that the function describing the relation between the conditional quantile and the explanatory variable is a polynomial function of a given degree, usually greater than 1. Thus, in the local polynomial-related methods the optimization problem takes the following form:

J. M. Uribe and M. Guillen, *Quantile Regression for Cross-Sectional and Time Series Data*, SpringerBriefs in Finance, https://doi.org/10.1007/978-3-030-44504-1_7

$$\hat{\alpha}(\theta, x) = \operatorname{argmin} \sum_{i=1}^{n} \rho_\theta \left(y_i - \alpha_0 - \alpha_1 (x_i - x) - \cdots - \frac{1}{p!} \alpha_p (x_i - x)^p \right). \tag{7.1}$$

The optimization problem above can be seen as an extension of a *weighted quantile regression* exercise in which the weighting is established using a *kernel* function and a *bandwidth* parameter h, which controls the degree of smoothness of the estimated function (Koenker 2005). The function **lprq** allows estimating a locally polynomial quantile regression in R.

The series method approach provides estimates of the conditional quantile function and its derivatives, based on series approximations to the nonparametric part of a partially linear quantile model. In this case, the optimization problem can be written as follows:

$$\hat{\alpha}(\theta) = \operatorname{argmin} \sum_{i=1}^{n} \rho_\theta \left(y_i - \sum_j \varphi_j (x_i) \alpha_j \right). \tag{7.2}$$

Based on the previous formulation, we are able to construct estimators of the linear functionals of the conditional quantile function by applying the plug-in principle to the series approximations (hence, the name of the method). A challenge to perform inference in this setting is that the dimension of the parameter space should increase with the sample size in order to reduce approximation error. Belloni et al. (2019) provide four methods that can be used to make inference on linear functionals of the conditional quantile function: pivotal, gradient bootstrap, Gaussian and weighted bootstrap. See Lipsitz et al. (2016) for the implementation of these four methods and point estimation in R using the **quantreg.nonpar** package.

Finally, we find a broad set of methods that can be encompassed under the term penalty methods. Unlike kernel methods, penalty methods rely on spline approximations to the nonparametric function of interest. Briefly, regression splines are obtained joining smoothed polynomial functions separated by a sequence of knots. That is, our optimization problem now turns into:

$$\hat{g}(\theta, x) = \operatorname{argmin} \sum_{i=1}^{n} \rho_\theta (y_i - g(x_i)) + \lambda P(g), \tag{7.3}$$

where $P(g)$ is the penalty function, which can take many possible shapes. Nevertheless, the total variation of the first derivative of $g(\cdot)$ is a particularly attractive option in the framework of quantile regression to define the shape of the penalty function. In this case, $P(g) = \mathcal{V}(g') = \int |g''(x)| dx$ and, naturally, λ controls the degree of smoothness of the function. Notice that as $\lambda \to \infty$, we constrain g to be closer to linear in x.

Solutions of the formulation in 16 with $P(g) = \mathcal{V}(g') = \int |g''(x)| dx$ are continuous and piecewise linear. One way to select λ is to minimize a Schwarz-type selection criterion. The function **rqss** can be used to implement penalty methods to estimate nonparametric quantile regressions in R.

7.2 The Cross-Quantilogram for Time Series

The cross-quantilogram was recently proposed by Han et al. (2016), as a way to measuring the dependence between two series at different conditional quantiles. They also developed a Box-Ljung Q-type statistic to test for directional predictability between two different series. The cross-quantilogram is a generalized version of the quantilogram statistic introduce to the literature by Linton and Whang (2007), which is univariate, while the cross-quantilogram is bivariate.

While in the traditional regression framework, the predictability or association between two-time series is measured from "mean" to "mean," both in the predictor and the predicted variables, a quantile regression generalizes this framework by allowing to quantify the association from the "mean" to "any given quantile" across the whole conditional distribution of the response variable. Let us say, the cross-quantilogram goes one step further by allowing to examine the dependence across two times series from "any given quantile" to "any given quantile" in the distribution of both the response and the explanatory variables. In this way, it is possible to study the association (or predictability) from the right tail of a variable to the left tail of the second variable, or from the median in one variable to the right tail in the second variable, and so for, with all possible combination of quantiles.

7.2.1 *The Cross-Quantilogram Definition*

Formally, consider $\{(\boldsymbol{y}_t, \boldsymbol{x}_t) : t \in \mathbb{Z}\}$ a stationary price time series where $\boldsymbol{y}_t = (y_{1t}, y_{2t})' \in \mathbb{R}^2$ and $\boldsymbol{x}_t = (x_{1t}, x_{2t}) \in \mathbb{R}^{d_1} \times \mathbb{R}^{d_2}$. Here, $x_{it} = \left[x_{it}^{(1)}, \ldots, x_{it}^{(d_i)}\right]' \in \mathbb{R}^{d_i}$ and $d_i \in \mathbb{N}$ for $i = 1, 2$. The formulae above means in words that x_{it} is an explanatory (or predictor) variable of y_{it}, and therefore the predictability is running from x to y. We can use $F_{y_i|x_i}(\cdot|x_{it})$ as to refer to the function of the series y_{it} given x_{it}, and $f_{y_i|x_i}(\cdot|x_{it})$ to denote the corresponding density function. We also can naturally define the conditional quantile function as $q_{it}(\tau_i) = \inf\{v\colon F_{y_i|x_i}(v|x_{it}) \geq \tau_i\}$ for $\tau_i \in (0, 1)$, and for $i = 1, 2$. Thereby, the cross-quantilogram is a measure of serial dependence between two events $\{y_{1t} \leq q_{1,t}(\tau_1)\}$ and $\{y_{2t-k} \leq q_{2t-k}(\tau_2)\}$ for any pair of $\tau = (\tau_1, \tau_2)'$ and for a given number k. One can think of the cross-quantilogram as the cross-correlation of the quantile-hit processes given by $\{1[y_{it} \leq q_{it}(\cdot)]\}$, and which is defined as follows:

$$\rho_\tau(k) = \frac{E\big[\psi_{\tau_1}(y_{1t} - q_{1t}(\tau_1))\psi_{\tau_2}(y_{2t-k} - q_{2t-k}(\tau_2))\big]}{\sqrt{E\big[\psi_{\tau_1}^2(y_{1t} - q_{1t}(\tau_1))\big]}\sqrt{E\big[\psi_{\tau_2}^2(y_{2t-k} - q_{2t-k}(\tau_2))\big]}}, \tag{7.4}$$

for $k = 0, \pm 1, \pm 2, \ldots$, where $\psi_a(u) \equiv 1[u < 0] - a$. Importantly, the cross-quantilogram is a statistic that is defined even in the case in which the two processes under analysis exhibit infinite moments, and it is also invariant to any strictly monotonic transformation, which is particularly convenient when working with prices or other series that require some sort of seasonal transformation, as its frequently the case in various applications of the energy field.

The empirical counterpart (in sample) of Eq. 17 relies on a linear conditional quantile regression as studies in the past chapters. Thus, it uses the following formula to estimate the conditional quantile functions, $\hat{q}_{it}(\tau_i)$:

$$\hat{\rho}_\tau(k) = \frac{\sum_{t=k+1}^{T} \psi_{\tau_1}(y_{1t} - \hat{q}_{1t}(\tau_1)) \psi_{\tau_2}(y_{2t-k} - \hat{q}_{2t-k}(\tau_2))}{\sqrt{\sum_{t=k+1}^{T} \psi_{\tau_1}^2(y_{1t} - \hat{q}_{1t}(\tau_1))} \sqrt{\sum_{t=k+1}^{T} \psi_{\tau_2}^2(y_{2t-k} - \hat{q}_{2t-k}(\tau_2))}}. \tag{7.5}$$

Notice that the cross-quantilogram takes into account dependence in terms of the direction of the deviation from the conditional quantiles and, as such, it measures the directional predictability from one series to another.

7.2.2 *Q-Test for Directional Predictability*

When one is interested in examining the null hypothesis $H_0 : \rho_\tau(1) = \ldots = \rho_\tau(p) = 0$ against the alternative hypothesis that $\rho_\tau(k) \neq 0$ for some $k \in \{1, \ldots, p\}$ the authors propose using a traditional Box-Pierce statistic for small samples (Box-Ljung) given by:

$$\hat{Q}_\tau^p \equiv T(T+2) \sum_{k=1}^{p} \hat{\rho}_\tau^2(k)/[T-k]. \tag{7.6}$$

7.2.3 *The Stationary Bootstrap*

In order to construct the null distribution of the cross-quantilogram and the Q-statistic presented above, it is convenient to use the stationary bootstrap developed by Politis and Romano (1994). The stationary bootstrap is a block bootstrap method with blocks of random lengths.

7.3 Quantile Regression Forests

The principle of quantile regression forests is the method of regression trees. When explaining a quantile of the response variable, the explanatory covariates confirm a tree-structure. This means that one covariate is used to form two groups, which correspond to two branches. In each branch, another covariate is used to divide the sample again. The sequence of tree spanning does not need to be the same in each of the branches. A forest is simply a collection of trees, where each tree has been obtained with a bootstrapped subsample of the original training sample.

There are several decisions that must be taken into account when generating trees. One of them is the size of the subsample to fit each tree; this is usually constant and equal to the initial training sample size, where bootstrap is implemented with replacement. When a tree node has less than a small number of observations, for instance, ten, then the node is not portioned any more.

The conditional distribution of the response variable is estimated by the weighted distribution suppose that we study Y given X = x based on the tree Ψ. Suppose the leaf which contains x then the weight $\omega i(x, \Psi)$ of observation for that leaf either 0 if it is not contained in the leaf or 1 divided by the number of observations in that leaf.

Let, the T trees of the random forests be $\Psi_1, \ldots, \Psi_T$, and $\omega i(x)$ be the average of $\omega i(x, \Psi)$ over all the trees. Then, the conditional quantile is the weighted empirical estimate.

Quantile regression forests are easy to implement via the existing R packages, but they are far less interpretable than classical quantile regressions. This is the reason why they are being used mostly for predictive purposes.

References

Belloni, A., Chernozhukov, V., Chetverikov, D., & Fernández-Val, I. (2019). Conditional quantile processes based on series or many regressors. *Journal of Econometrics, 213*(1), 4–29.

Cai, Z. (2002). Regression quantiles for time series. *Econometric Theory, 18*(1), 169–192.

Han, H., Linton, O., Oka, T., & Whang, Y. J. (2016). The cross-quantilogram: Measuring quantile dependence and testing directional predictability between time series. *Journal of Econometrics, 193,* 251–270.

Koenker, R. W. (2005). *Quantile regression.* Cambridge: Cambridge University Press.

Linton, O., & Whang, Y. J. (2007). The quantilogram: With an application to evaluating directional predictability. *Journal of Econometrics, 141,* 250–282.

Lipsitz, M., Belloni, A., Chernozhukov, V., & Fernández-Val, I. (2016). quantreg.nonpar: An R package for performing nonparametric series quantile regression. *The R Journal, 8*(2), 370–381.

Politis, D. N., & Romano, J. P. (1994). The stationary bootstrap. *Journal of the American Statistical Association, 89,* 1303–1313.

Chapter 8
What Have We Learned from Quantile Regression? Implications for Economics and Finance

Abstract In this chapter, we conclude and we emphasize that quantile regression is suitable for predictive modeling when the response is asymmetric or non-normally distributed conditional on the covariates. We note that usually, the model aims at the lower or upper percentiles of the response.

Keywords Extremes · Percentile · Prediction · Conditional distribution · Non-normality · Asymmetry

In recent years, there is a booming of interest in the analysis of risk in all domains of economics and finance. This is so because most policy decisions are aimed at scenarios that can be catastrophic. Then, noncentral scenarios and extreme quantiles become of primary interest.

We believe that the quantile regression offers a unique approach that can be helpful in many areas of economics and finance by providing a way to predict the conditional distribution of a dependent variable.

By looking at quantiles, especially at low percentiles (5% or 10%) or at high percentiles (90, 95%), we can predict the scope of a dependent variable. This is like its major domain, or the interval that is expected to contain the majority of values given the information on the covariates. This is done in such a way that there is no assumption of normality, which assumes that these quantiles are located symmetrically around the mean of the conditional distribution.

J. M. Uribe and M. Guillen, *Quantile Regression for Cross-Sectional and Time Series Data*, SpringerBriefs in Finance, https://doi.org/10.1007/978-3-030-44504-1_8

Appendix
Programs for Quantile Regression and Implementation in R

The data necessary to carry out the estimations in this book are available from the authors and are of public access.

```
#==============================================================
# Chapter 4: Cross-sectional quantile regression
##=============================================================

### Chapter 4 ### Cross-sectional quantile regression
### Importing the data set ###

## clear all existing objects
rm(list=ls(all=TRUE))

## set working directory
setwd("introduce your directory here")

## obtain data
load("RECS.Rdata")
attach(RECS)

# Figure 2 #############################################
# Quantiles for Cost of Electricity (US$) from RECS in 2015
#########################################################

x <- DOLLAREL
n <- length(DOLLAREL)
plot((1:n - 1)/(n - 1), sort(x), type="l",
xlab = "Fraction of data",
ylab = "Quantiles of the total cost of electricity (US$)")

# Figure 3 #############################################
# OLS regression of Energy Consumption on Square
Footage Unweighted and using the sample weights
##########################################################

# defining weights

RECSnodes <- svydesign(id=~1, weights=~1, data=RECS)
svymean(~DOLLAREL+TOTSQFT_EN, RECSnodes)
```


J. M. Uribe and M. Guillen, *Quantile Regression for Cross-Sectional and Time Series Data*, SpringerBriefs in Finance, https://doi.org/10.1007/978-3-030-44504-1

```
# without weights

RECSdes <- svydesign(id=~1, weights=~NWEIGHT, data=RECS)
svymean(~DOLLAREL+TOTSQFT_EN, RECSdes)

# Linear regression using lm and plotting of the results

OLS_RECSunw<- lm(DOLLAREL~TOTSQFT_EN, data=RECS, model=T)
summary(OLS_RECSunw)

OLS_RECSw<- lm(DOLLAREL~TOTSQFT_EN, data=RECS, model=T,
weights=NWEIGHT)
summary(OLS_RECSw)

# Top panel

plot(DOLLAREL ~ TOTSQFT_EN, data = RECS, cex=0.1, xlab =
"Surface (squared feet)", ylab =
"Electricity Energy Consumption in 2015 ($)")
abline(coef(OLS_RECSunw)[1:2])
cf <- round(coef(OLS_RECSunw), 3)
eq <- paste0("Cost= ", cf[1],
ifelse(sign(cf[2])==1, " + ", " - "), abs(cf[2]), " Surface")
mtext(eq, 3, line=-2)

# Bottom panel

plot(RECS$DOLLAREL ~ RECS$TOTSQFT_EN, data = RECS,
cex=NWEIGHT*2/max(NWEIGHT),
xlab = "Surface (squared feet)", ylab =
"Electricity Energy Consumption in 2015 ($)")
abline(coef(OLS_RECSw)[1:2])
cf <- round(coef(OLS_RECSw), 3)
eq <- paste0("(Weighted OLS) Cost= ", cf[1],
ifelse(sign(cf[2])==1, " + ", " - "), abs(cf[2]), " Surface")
mtext(eq, 3, line=-2)

# Figure 4 and Table 2 ###################################
# Weighted OLS and weighted quantile regressions' coefficients
##################################################
#### The following command calls the package 'quantreg'
available in R to perform the estimations and plot the figure.

library(quantreg)

# Quantile regression fitting

QR_RECS2 <- rq(DOLLAREL~TOTSQFT_EN, data =RECS, model=TRUE,
weights=NWEIGHT, tau= seq(0.05,0.95,by=0.10))
QR.1<-summary(QR_RECS2)

# Potting of the results

plot(QR.1, main=c("Intercept","Total square footage"))

# Figure 5 ###############################################
## Plot of the simple quantile regressions for the Energy
Consumption as a function of the household Surface
```

```
plot(RECS$TOTSQFT_EN, RECS$DOLLAREL, xlab = "Surface (squared
feet)", ylab =
"Electricity Energy Consumption in 2015 ($)")

# define the quantile levels and employ them in the regression

u = seq(.05,.95,by=.01)
length(u)
coefstd = function(u) summary(rq(DOLLAREL~TOTSQFT_EN, data =
RECS , weights=NWEIGHT,tau=u))$coefficients[,2]
coefest = function(u) summary(rq(DOLLAREL~TOTSQFT_EN, data =
RECS , weights=NWEIGHT,tau=u))$coefficients[,1]
CS = Vectorize(coefstd)(u)
CE = Vectorize(coefest)(u)
abline(a=CE[1,1], b=CE[2,1],col="blue")
abline(a=CE[1,45], b=CE[2,45], col="gray")
abline(a=CE[1,91], b=CE[2,91], col="red")
text(locator(), labels = c("5%", "50%", "95%"))

# Table 3 ##########################################
# Weighted quantile regressions' coefficients at different
quantiles

QR_RECS3.1 <-rq(DOLLAREL~factor(SWIMPOOL)+TOTSQFT_EN+TOTCSQFT
+TOTHSQFT+BEDROOMS+factor(SOLAR)+TVCOLOR+NUMSMPHONE+NUMFLOORFAN
+NHSLDMEM,
data =subset(RECS,SWIMPOOL>=0) , weights=NWEIGHT, model=TRUE,
tau= 0.25)
summary.rq(QR_RECS3.1, se="rank",covariance = T)

QR_RECS3.2 <- rq(DOLLAREL~factor(SWIMPOOL)+TOTSQFT_EN+TOTCSQFT
+TOTHSQFT+BEDROOMS+factor(SOLAR)+TVCOLOR+NUMSMPHONE+NUMFLOORFAN
+NHSLDMEM,
data =subset(RECS,SWIMPOOL>=0) , weights=NWEIGHT, model=TRUE,
tau=0.50)
summary.rq(QR_RECS3.2, se="rank",covariance = T)

QR_RECS3.3 <- rq(DOLLAREL~factor(SWIMPOOL)+TOTSQFT_EN+TOTCSQFT
+TOTHSQFT+BEDROOMS+factor(SOLAR)+TVCOLOR+NUMSMPHONE+NUMFLOORFAN
+NHSLDMEM,
data =subset(RECS,SWIMPOOL>=0) , weights=NWEIGHT, model=TRUE,
tau= 0.75)
summary.rq(QR_RECS3.3, se="rank",covariance = T)

QR_RECS3.4 <- rq(DOLLAREL~factor(SWIMPOOL)+TOTSQFT_EN+TOTCSQFT
+TOTHSQFT+BEDROOMS+factor(SOLAR)+TVCOLOR+NUMSMPHONE+NUMFLOORFAN
+NHSLDMEM,
data =subset(RECS,SWIMPOOL>=0) , weights=NWEIGHT, model=TRUE,
tau= 0.95)
summary.rq(QR_RECS3.4, se="rank",covariance = T)

# Figure 6 ##############################################
# Plot of the coefficients of the weighted quantile regression
model and the OLS model for Electricity Consumption

QR_RECS4 <- rq(DOLLAREL~factor(SWIMPOOL)+TOTSQFT_EN+TOTCSQFT
+TOTHSQFT+BEDROOMS+factor(SOLAR)+TVCOLOR+NUMSMPHONE+NUMFLOORFAN
```

```
+NHSLDMEM,
data =subset(RECS,SWIMPOOL>=0) , weights=NWEIGHT, model=TRUE,
tau= seq(0.05,0.95,by=0.10))
QR.2<-summary(QR_RECS4)

plot(QR.2, main=c("Intercept","Swimming Pool","Square Footage",
"Cooled square footage","Heated square footage",
"Num.Bedrooms","Electricity generation from solar",
"TVCOLOR", "Smart Phones", "Floor fans",
"Num. of household members"))

# OLS regression
OLS_RECS4 <- lm(DOLLAREL~factor(SWIMPOOL)+TOTSQFT_EN+TOTCSQFT
+TOTHSQFT+BEDROOMS+factor(SOLAR)+TVCOLOR+NUMSMPHONE+NUMFLOORFAN
+NHSLDMEM,
data =subset(RECS,SWIMPOOL>=0) , weights=NWEIGHT, model=TRUE)
summary(OLS_RECS4)
confint(OLS_RECS4,level = 0.95)

# Table 4 ################################################
## Weighted OLS and unweighted unconditional quantile
regressions' coefficients at different quantiles
## the following command calls the package 'uqr' available in R
to conduct unconditional quantile regressions

library(uqr)

formula = DOLLAREL~TOTSQFT_EN
UQR.1=urq(formula,data = RECS, kernel="gaussian", tau=0.25)
UQR.1$coefficients
summary=urqCI(urq =UQR.1,R = 100,graph = TRUE,seed = 1234)
summary$results
UQR.2=urq(formula,data = RECS, kernel="gaussian", tau=0.5)
UQR.2$coefficients
summary=urqCI(urq =UQR.2,R = 100,graph = TRUE,seed = 1234)
summary$results
UQR.3=urq(formula,data = RECS, kernel="gaussian", tau=0.75)
UQR.3$coefficients
summary=urqCI(urq =UQR.3,R = 100,graph = TRUE,seed = 1234)
summary$results
UQR.4=urq(formula,data = RECS, kernel="gaussian", tau=0.95)
UQR.4$coefficients
summary=urqCI(urq =UQR.4,R = 100,graph = TRUE,seed = 1234)
summary$results

#===========================================================
# Chapter 5: Time series quantile regression
##==========================================================
## clear all existing objects
rm(list=ls(all=TRUE))

## set working directory
setwd("introduce here your working directory")

##======================
## data
```

```
##========================

load("PJM.RData")

data=data[,2:3]
t=dim(data)[1]
n=dim(data)[2]
price=matrix(,t,n)
for (i in 1:n)
{price_=as.numeric(data[,i])
price[,i]=price_}
power=price[,1]
gas=price[,2]
colnames(price) <-c("power", "gas")

## Define data as to explain power prices using gas lags
DATA = cbind(power, gas)

#Variable Definition
Y=power[1:(t-5)]
X=gas[1:(t-5)]
X1=gas[2:(t-4)]
X2=gas[3:(t-3)]
X3=gas[4:(t-2)]
X4=gas[5:(t-1)]
X5=gas[6:t]
Y1=power[2:(t-4)]
Y2=power[3:(t-3)]
Y3=power[4:(t-2)]
Y4=power[5:(t-1)]
Y5=power[6:t]

# Figure 7 ################################################
# Quantiles for the Price of Electricity (US$) EIA ###

x <- Y
n <- length(x)
plot((1:n - 1)/(n - 1), sort(x), type="l",
xlab = "Fraction of data",
ylab = "Quantiles of the price of electricity (US$)")

#Figure 8 ##################################################
# OLS regression of electricity prices natural gas prices
#########################################################

#OLS Regression
olsreg <- lm(Y~X)
summary(olsreg)

plot(Y ~ X, cex=0.1, xlab = "Natural gas prices ($)", ylab =
"Electricity prices ($)")
abline(coef(olsreg)[1:2])
cf <- round(coef(olsreg), 3)
eq <- paste0("electricity= ", cf[1],
ifelse(sign(cf[2])==1, " + ", " - "), abs(cf[2]), "gas")
mtext(eq, 3, line=-2)
```

```
mean(Y)
sqrt(var(Y))
mean(X)
sqrt(var(X))

summary(X)
summary(Y)

#Figure 9 and Table 6 ####################################
# OLS and quantile regression coefficients at different
quantiles ##################################################
#### The following command calls the package ???quantreg???
available in R to perform the estimations and plot the figure.

library(quantreg)

# Quantile Regression
Electricity=Y
Gas=X
quantreg <- rq(Electricity~Gas, tau=seq(0.1,0.9,0.1))
summary(quantreg)
quantreg.plot=summary(quantreg)
plot(quantreg.plot, main=c("Intercept","Natural gas prices"))

### Figure 10 ##############################################
### Quantile regressions for the Electricity prices as a
function of the natural gas prices ##################

u = seq(.05,.95,by=.01)
length(u)
coefstd = function(u) summary(rq(Electricity~Gas
,tau=u))$coefficients[,2]
coefest = function(u) summary(rq(Electricity~Gas
,tau=u))$coefficients[,1]
CS = Vectorize(coefstd)(u)
CE = Vectorize(coefest)(u)
plot(Electricity~ Gas,cex=0.1, xlab = "Natural gas ($)", ylab
=
"Electricity ($)")
abline(a=CE[1,1], b=CE[2,1],col="blue")
abline(a=CE[1,45], b=CE[2,45], col="gray")
abline(a=CE[1,91], b=CE[2,91], col="red")
#text(locator(), labels = c("5%", "50%", "95%"))

tau=seq(0.1,0.9,0.1)

### Table 7 ################################################
### AIC criteria for quantile models with autorregressions
##############################################
n_=length(Y)
l=length(tau)
#AR 1
k=1
quantreg1 <- rq(Y~Y1, tau=seq(0.1,0.9,0.1))
RSS_1=quantreg1$residuals^2
AIC1= 2*k+(2*n_*log(colSums(RSS_1)))
```

```
#AR 2
k=2
quantreg2 <- rq(Y~Y1+Y2, tau=seq(0.1,0.9,0.1))
RSS_2=quantreg2$residuals^2
AIC2= 2*k+(2*n_*log(colSums(RSS_2)))

#AR 3
k=3
quantreg3 <- rq(Y~Y1+Y2+Y3, tau=seq(0.1,0.9,0.1))
RSS_3=quantreg3$residuals^2
AIC3= 2*k+(2*n_*log(colSums(RSS_3)))

#AR 4
k=4
quantreg4 <- rq(Y~Y1+Y2+Y3+Y4, tau=seq(0.1,0.9,0.1))
RSS_4=quantreg4$residuals^2
AIC4= 2*k+(2*n_*log(colSums(RSS_4)))

#AR 5
k=5
quantreg5<- rq(Y~Y1+Y2+Y3+Y4+Y5, tau=seq(0.1,0.9,0.1))
RSS_5=quantreg5$residuals^2
AIC5= 2*k+(2*n_*log(colSums(RSS_5)))

AIC=matrix(c(AIC1,AIC2,AIC3,AIC4,AIC5),,1)

### Figure 11 ##############################################
### Plot of the intercept and the coefficients associated
to different lags of the price of electricity at different
quantiles ########################

summary(quantreg5)
quantreg5.plot=summary(quantreg5)
plot(quantreg5.plot, main=c("Intercept","Lag 1","Lag 2","Lag
3","Lag 4","Lag 5"))

### Figure 12 ##############################################
### Plot of electricity prices and percentiles 10th (left)
and 90th (right), including five lags of prices as explanatory
variables ####################################

y2=quantreg5$fitted.values[,9]
y3=quantreg5$fitted.values[,1]
x = seq(1,941,by=1)
plot(x,Electricity, xlab='Observation', ylab='Return')
lines(x,y3,col="red")

plot(x,Electricity, xlab='Observation', ylab='Return')
lines(x,y2,col="blue")
```

The manufacturer's authorised representative in the EU is Springer Nature Customer Service Centre GmbH, Europaplatz 3, 69115 Heidelberg, Germany. If you have any concerns regarding our products, please contact ProductSafety@springernature.com

Printed and bound by CPI Group (UK) Ltd, Croydon, CR0 4YY
05/10/2026
02228152-0002